Principles of Industrial Safety

Joel M. Haight, Editor

Copyright © 2012 by the American Society of Safety Professionals
All rights reserved.

Copyright, Waiver of First Sale Doctrine
All rights reserved. No part of this work may be reproduced or transmitted in any form or by any means, electronic or mechanical for commercial purposes, without the permission in writing from the Publisher. All requests for permission to reproduce material from this work should be directed to: The American Society of Safety Professionals, ATTN: Manager of Technical Publications, 520 N. Northwest Highway, Park Ridge, IL 60068.

Disclaimer
While the publisher and authors have used their best efforts in preparing this book, they make no representations or warranties with respect to the accuracy or completeness of the contents, and specifically disclaim any implied warranties of fitness for a particular purpose. The information herein is provided with the understanding that the authors are not hereby engaged in rendering professional or legal services. The mention of any specific products herein does not constitute an endorsement or recommendation by the American Society of Safety Professionals, and was done solely at the discretion of the author. Note that any excerpt from a National Fire Protection Association Standard reprinted herein is not the complete and official position of the NFPA on the referenced subject(s), which is represented only by the cited Standard in its entirety.

Library of Congress Cataloging-in-Publication Data

Principles of industrial safety / Joel M. Haight, editor.
    pages cm
 Includes bibliographical references and index.
 ISBN 978-1-885581-75-4 (alk. paper)
1. Industrial safety.
 T55.P753 2013
 658.4'08--dc23
                                2012043234

Notice of Name Change
The American Society of Safety Engineers (ASSE) is now the American Society of Safety Professionals (ASSP).

Managing Editor: Michael F. Burditt, ASSP
Editor: Jeri Ann Stucka, ASSP
Text design and composition: Cathy Lombardi
Cover design: Image Graphics

Printed in the United States of America

24 23 22 21 20 19   5 6 7 8 9 10 11 12

PRINCIPLES OF INDUSTRIAL SAFETY

Contents

Foreword	iv
About the Editor and Authors	viii
Chapter 1: Regulatory Issues Anjan K. Majumder	1
Chapter 2: General Safety Management Jeffery C. Camplin	15
Chapter 3: Managing a Safety Engineering Project Joel M. Haight	65
Chapter 4: Global Issues Kathy A. Seabrook	101
Chapter 5: Cost Analysis and Budgeting T. Michael Toole	129
Chapter 6: Benchmarking and Performance Criteria Christopher Janicak	151
Chapter 7: Best Practices Linda Rowley	173
Index	187

Foreword

THE FIRST CHAPTER in this book is a summary of some of the regulations and voluntary and consensus standards associated with managing safety engineering work. It does not include everything and every detail but provides very useful guidance of what to look for while doing safety engineering and management of workplaces. For example, it talks about ISO (International Organization for Standardization) 9001 and 14001 but not OHSAS 18001, which is an internationally recognized occupational, health and safety management system series standard.

Excellence in safety engineering and management of workplaces is based on management leadership and employee involvement, which together successfully address worksite analysis; hazard prevention and control; training; and continuous improvement. This chapter helps safety professionals understand what requirements (rules and regulations) must be addressed, what consensus standards should be considered and, lastly, what other things (best practices) would add value in their worksite analysis to plan for hazard prevention and control. It also discusses executive orders and how the federal agencies may be influenced or affected by them; what Code of Federal Register (CFR) applies to which agency; and what voluntary programs can be adopted by workplaces to take proactive approach towards the environmental and occupational safety and health issues. The somewhat misunderstood method by which federal agencies, through a process called "incorporation by reference" is explained, as is the practice by authorities having jurisdiction (AHJ) of making voluntary consensus standards mandatory by adoption.

The author, Anjan Majumder, provides an overview of OSHA consultative recognition programs, before discussing the federal statutes which provide the legal basis for the EPA. ISO standards are also introduced and their role in international trade briefly noted.

In the chapter which follows, "General Safety Management," Jeff Camplin begins by discussing the definition of an effective safety and health management system. Ownership of a company's safety management system by senior management is an outgrowth of recognizing the critical link between safety and productivity and profitability.

The role of a company's business culture in integrating safety with all facets of business operations is addressed as well. Camplin also notes the role that consensus standards play in providing basic requirements for occupational health and safety management systems. ANSI Z10, *Standard for Occupational Health and Safety Management Systems,* is an example of a consensus standard that provides basic requirements for occupational health and safety management systems rather than detailed requirements, in tacit recognition that one size does not fit all when it comes to safety management systems.

The discussion moves on to OSHA's management approach to building an effective safety management system. The author next looks at each of the four elements of the OSHA system and other resources available from OSHA Some tools of the trade are provided to assist the SHE professional in the implementation of an effective safety management system.

Managing safety engineering work or the safety engineering aspects of any work or project involves many of the same management tools and applications that any other type of work or project would require. The third chapter is dedicated to helping readers to not only learn some of these basic tools and applications, if they haven't seen them before or are not familiar with them, but it is also dedicated to showing the readers the relationships and similarities of their use on other types of projects. This chapter provides a basis and some background on what is meant by "a safety engineering project" and covers the basics of what tools and techniques are required to help to ensure the work is done to achieve the organization's objectives in the time intended and at the expected cost. In the case of safety engineering work, of course, the project's objectives also have to be met without any injuries or other undesired incidents.

In managing any organized project, a project manager should be concerned about and address issues such as developing a project plan, determining a budget and a schedule, sorting out equipment availability, and determining the efficient allocation of human resources in the form of numbers and skill sets or disciplines. The chapter begins with some foundation information about safety engineering and moves into the development of a project plan in terms of work and task breakdown and allocation. Two of the main concerns of any project manager are the project's schedule and the project's budget. This chapter addresses both, and it includes solid, real-life examples of each. For scheduling the author addresses two frequently used techniques in relatively significant depth. Techniques such as Gantt charting and Project Evaluation and Review Technique (PERT) charting are presented, each with an in-depth example. To address budgeting, the chapter includes basic engineering economics and decision-making where an example of a net present value (NPV) determination is made for a real-life example decision between upgrading an existing facility vs. building a new facility. General budgeting through identifying and organizing required tasks, as well as methods for cost tracking are presented, again with detailed examples.

The reader will be able to find coverage of concepts and information that support project management, such as team building, leadership, people development, human resource allocation using efficiency models, conflict resolution, learning curves, work sampling and its associated uncertainty calculations, performance ratings and time study, etc.

While this chapter, as is the case with all chapters in this handbook, is a summary of a much larger body of literature, the reader will be able to find thorough and detailed coverage of each of the topics. Given that this is a handbook, readers are encouraged to seek greater depth in this larger body of knowledge. The reference list at the end of the chapter provides some of the resources to which the reader is referred. The chapter contains modified versions of real cases, analyses and exercises that the author actually conducted in real settings. The final case study and exercises for the chapter are also detailed and were developed from a real case in a real industrial setting.

With ever-increasing globalization, safety and health professionals who seek to understand the business environment in which they operate are better able to anticipate changes that will impact workplace safety and health within their own organizations in the U.S. and around the world. This is essence of Chapter 4, "Global Issues." Developing your "global fluency" and strategic business perspective, in addition to your SH&E management systems and technical safety and health skills, are keys to your success in the safety and health profession. In today's global economy, the likelihood is increasing you will work with colleagues from other countries. It could be as a SHE auditor auditing a site outside your home country, working on a global SHE project team (in person or via video conferencing) or possibly relocating as an expatriate to another country to work with local nationals to transfer skills and knowledge on safety and health.

According to the United States (U.S.) Department of Commerce, foreign direct investment in the US was $194 billion in 2010 and $1.7 trillion from 2000–2012. Over the same 10-year period, there were 2 million manufacturing jobs created in the U.S. from these non-U.S.-owned companies; on average, their employees received 30% higher pay than their counterparts in U.S.-owned companies.[1] For the safety, health and environmental professional, this means over the course of your career, the likelihood you will work for a non-U.S. company while domiciled in the U.S. continues to increase. That may require an understanding of new cultural norms, implementing company safety and health standards and aligning business culture globally. Working in U.S.-based multinational companies will also require similar skills and knowledge as you manage, across time zones and cultures, to be an effective business leader.

This chapter will provide strategies for managing safety and health risk in a multinational organization. The principles of global safety and health management are scalable to most sized organizations, and the chapter reviews some of the basics you will need to develop your own "global fluency" and an understanding of global safety and health management principles.

Note: While environmental management is not within the scope of this chapter, the reader should keep in mind the principles of managing environmental impacts are aligned with those of safety and health management outlined in this chapter. An expectation of a certain level of environmental management systems knowledge is not uncommon in many companies.

The last three chapters focus specifically on management issues. As Dr. Toole points out in Chapter 5, "Cost Analysis and Budgeting,"

> Although safety programs are rarely viewed as increasing revenues, they are typically recognized as essential to reducing costs, or to ensuring that costs do not increase faster than revenues do. Effective safety programs, therefore, can contribute to achieving corporate goals involving profitability just as much as more visible initiatives that involve new products, services, and so on.

This chapter will provide the reader with tools that can objectively evaluate whether a project should be pursued. While the tools are often referred to as associated with *engineering economics*, they are really financial analysis tools that are applied every day to projects having nothing to do with either engineering or safety. For example, the tools summarized in this chapter can help one decide whether to invest in real estate, launch a new product, or buy an extended warranty on a consumer purchase. Similarly, the fundamental concepts from engineering economics introduced at the beginning of this chapter apply to all financial analysis situations, not just to those involving safety or engineering.

The next chapter, "Benchmarking and Performance Criteria," is written based on the author's belief that "Safety performance should be evaluated in an organization in the same manner as productivity and other aspects of business." The author, Dr. Christopher Janicak, emphasizes the importance of a structural approach to key performance improvement measures: planning, establishing goals and objectives, identifying valid measures, conducting proper data analysis, and implementing appropriate follow-up measures. Valid measures are true indicators of performance. Likewise, corrective actions are also valid means for improving performance. Leading, trailing, and current measurement indicators are discussed.

Critical aspects of benchmarking a company's safety performance is discussed next, followed by a discussion of how quality control techniques can be applied to improving safety and health programs.

The best practices in managing safety engineering work reach beyond using safety engineering techniques, instituting a risk management model, generating a project management plan, or controlling costs. These are important elements in system safety, but an ideal system safety program is an organizational approach that enables safety engineering work. Safety engineering work can be significantly affected by the management structure. The management structure and organizational culture greatly influence the degree of support and

---

[1] Payne, David and Yu, Fenwick. June 2011. ESA Issue Brief #02–11. "Foreign Direct Investment in the United States." U.S. Department of Commerce, Economics and Statistics Administration. Accessed July 15, 2012. http://www.esa.doc.gov/sites/default/files/reports/documents/fdiesaissuebriefno2061411final.pdf

collaboration in system safety, communications and information flow, and responsibility and accountability. Fostering a positive safety culture encourages participation and feedback concerning a system, which can bring new insights and identify emerging hazards.

Leaders and managers routinely make risk decisions, recognizing that risk and uncertainty are essential to business and growth. Without risk, the world is static and progress is stifled; however, too much risk can be little more than a gamble. System safety is a risk-based approach of anticipating and identifying existing and emerging hazards that may occur during the life cycle of a system. It is unrealistic to assume that: 1) every hazard can be anticipated and identified; 2) every hazard must be engineered out of the system; and 3) other hazards, in addition to those previously identified during the design and test phases, will not occur during the life cycle of a system.

In industrial accident prevention programs, greater focus is on human errors and unsafe conditions that could cause an accident. In safety engineering work, the focus is more on identifying hazards in hardware, software, and operational procedures that could cause harm to the system or operations. Both perspectives are important in system safety. An overarching component, the management framework, takes into consideration the direct, as well as the indirect, factors that can affect a system.

To some extent, the system safety program is affected by government, technology, and the environment in which the organization is operating. Indirect influences such as the political, sociological, and organizational changes can affect safety engineering work. For example, a catastrophic accident often brings about national attention and pressure to find engineering solutions to "fix" a problem. A structured management process ensures a thorough and orderly process for performing safety engineering work.

The purpose of this chapter is not to identify the best safety engineering methodology, or to imply that a management structure is the key to identifying and eliminating all hazards within a system. According to Cantrell and Clemens, "…it remains true that no single technique exists for identifying system hazards capable of finding them all, and no method for verifying that it might have done so."[2] The best safety engineering techniques are those that are most suited to identifying hazards. The best practice in managing safety engineering work is to employ a risk management system that promotes a positive safety culture.

---

[2]*Professional Safety*, November 2009, p. 35.

ABOUT THE EDITOR

In 2009, Joel M. Haight, Ph.D., P.E., was named Branch Chief of the Human Factors Branch at the Centers for Disease Control and Prevention (CDC)—National Institute of Occupational Safety and Health (NIOSH) at their Pittsburgh Office of Mine Safety and Health Research. He continues in this role. In 2000, Dr. Haight received a faculty appointment and served as Associate Professor of Energy and Mineral Engineering at the Pennsylvania State University. He also worked as a manager and engineer for the Chevron Corporation domestically and internationally for eighteen years prior to joining the faculty at Penn State. He has a Ph.D. (1999) and Master's degree (1994) in Industrial and System Engineering, both from Auburn University. Dr. Haight does human error, process optimization, and intervention effectiveness research. He is a professional member of the American Society of Safety Engineers (where he serves as Federal Liaison to the Board of Trustees and the ASSE Foundation Research Committee Chair), the American Industrial Hygiene Association (AIHA), and the Human Factors and Ergonomics Society (HFES). He has published more than 30 peer-reviewed scientific journal articles and book chapters and is a co-author and the editor-in-chief of ASSE's *The Safety Professionals Handbook* and the John Wiley and Sons, *Handbook of Loss Prevention Engineering*.

ABOUT THE AUTHORS

Jeffery C. Camplin, M.S., CSP, CPEA, is President of Camplin Environmental Services, Inc., a safety and environmental consulting firm he founded in 1991.

Christopher A. Janicak, Ph.D., CSP, CEA, ARM, is Professor and Doctoral Program Coordinator for the Department of Safety Sciences at Indiana University of Pennsylvania (IUP).

Anjan K. Majumder, M.S., CHMM, EIT, CSP, is a Safety Engineer with the Federal Aviation Administration.

Linda S. Rowley, M.B.A., CSP, is the Safety and Health Manager for the U.S. Bureau of Reclamation.

Kathy A. Seabrook, CSP, CMIOSH (UK), is President of Global Solutions, Inc., which she founded in 1996. She holds professional safety and health certifications in the US and UK and graduated with a B.S. in Chemistry from James Madison University. She is a published author in the field of global safety, health, and environmental management, a member of the ANSI Z10 Standards Development Committee, and served on the working party for OHSAS 18002. Kathy has been a member of the ASSE National Faculty since 1996, teaching Global SHE Management and Sustainability & the Safety Professional. She currently serves on the ASSE Board of Directors and is an advisory committee member to the ASSE International Practice Specialty and Women in Safety Engineering, as well as the British Institution of Occupational Safety and Health's International Group.

T. Michael Toole, Ph.D., P.E., is Associate Professor of Civil and Environmental Engineering at Bucknell University.

About the Editor

REGULATORY ISSUES

Anjan K. Majumder

1

LEARNING OBJECTIVES

- Be able to define safety engineering management.

- Understand the purpose and goal of safety engineering and management.

- Be aware of regulations applicable to workplaces: mandatory requirements, consensus standards, codes and standards (including international standards), best management practices adopted by private or public entities, occupational safety and health standards applicable to private industry as well as federal agencies (including environmental laws and regulations).

- Explain the two main functions of OSHA under the Occupational Safety and Health Act (OSH Act).

- Be aware of voluntary programs that help businesses and organizations to proactively attain and maintain safe work environments.

THIS CHAPTER IS a summary of some of the regulations and voluntary and consensus standards associated with managing safety engineering work. It is not an exhaustive list, nor does it contain exact wording from the regulations. It is meant to inform the reader of some of the more important regulations and standards that guide safety engineering work and to provide the reader with a pathway to find this regulatory and consensus information. It is expected that once the reader determines the need to gather regulatory information, he or she will refer to the regulations and standards themselves. This chapter provides an informative background and summary upon which the reader can build a foundation of regulatory and consensus standards' knowledge.

SAFETY ENGINEERING AND MANAGEMENT

Safety engineering management is a part of the general management system. It is a combination of risk identification, risk mitigation, and balancing among levels of acceptable risk, cost, and availability of technology. Safety engineering management includes system safety, occupational safety in the workplace, and behavioral safety. It requires implementation of regulatory requirements, consensus standards, and best management practices in the policies and procedures of safety engineering management. The regulatory requirement is the mandatory part, while the level and extent of consensus standards or best management practices can vary according to company policies.

A goal of safety engineering management is the efficient implementation of safety, reliability, quality, and cost-effectiveness. Some compliance policies applicable to specific establishments may be more stringent than others, based on the requirements of a federal agency, state or local authority having jurisdiction (AHJ), or company policy. This chapter discusses government regulations as well as codes and standards that can become mandatory for a company, agency, or AHJ if their policy has been established that way. By the end of the chapter, the reader should be familiar with the laws, regulations, codes, and standards that might need to be applied in any workplace when implementing safety engineering management.

Requirements that may apply to safety engineering and management are of the following types:
1. Executive orders
2. *Code of Federal Regulations* (CFR)
3. Federal agency orders and standards
4. State and local regulations
5. Consensus standards
6. International Organization for Standardization (ISO) standards
7. Best management practices

Executive Orders

An *executive order* is a formal or authoritative proclamation issued by a member of the executive branch of government, usually the head of the branch. Most executive orders are issued by the president of the United States. For example, Executive Order 12196 of February 26, 1980, Occupational Safety and Health Programs for Federal Employees (Federal Register 1980), orders all heads of agencies to furnish to employees places and conditions of employment that are free from recognized hazards that are causing or are likely to cause death or serious physical harm. Executive Order 13148, of April 21, 2000, Part 4, Sec 401 (EPA 2000), states that each federal agency, within eighteen months of the date of that order, must conduct an agency-level environmental management system (EMS) self-assessment—based on the Code of Environmental Management Principles for Federal Agencies, which has been developed by the Environmental Protection Agency (EPA) or another appropriate EMS framework.

Executive orders are found in sequential editions of Title 3 of the CFR. Each executive order is signed by the president of the United States and received by the Office of the Federal Register. The National Archives and Records Administration (NARA) of the Office of the Federal Register publishes the *Federal Register*, an official daily publication containing rules, proposed rules, and any notices created by federal agencies and organizations, as well as executive orders and other presidential documents.

Code of Federal Regulations (CFR)

The CFR is an annual codification of general and permanent rules. It is divided into 50 titles representing the broad areas that are subject to federal regulation. Each title is divided into chapters that are assigned to agencies issuing regulations pertaining to that broad subject area. Each chapter is divided into parts, and each part is divided into sections. The CFR provides the official and complete text of agency regulations in one organized publication. Each volume of the CFR is updated once each calendar year and and is issued quarterly (for example, the volume containing titles 1–16 is issued on January 1, the volume containing titles 17–27 is issued on April 1, and so on). The CFR is available electronically (free of charge) as well as in paper publication (by full set subscription or individual copy purchase). It can be obtained from the superintendent of documents or at any federal depository library. Among the 50 titles, some of the following may be required to deal with issues of safety engineering and management in some work environments.

Title 10: Energy
Title 21: Food and Drugs
Title 23: Highways
Title 29: Labor
Title 40: Protection of Environment
Title 42: Public Health
Title 44: Emergency Management and Assistance
Title 49: Transportation

Federal Agency Orders and Standards

Apart from executive orders and federal regulations, federal agencies have their own orders and standards.

All federal agencies have developed orders and standards for their own agency, but some have orders or standards that are applicable to other industries or workplaces as well. For example, the Department of Energy (DOE) has orders and technical standards that are applicable to DOE facilities, but the Department of Transportation's (DOT) classifications and placards for dangerous goods are applicable to any company that transports items by air, water, rail, public road, or private vehicle that qualifies under DOT regulations. The following government agencies have safety and health regulations that may be applicable to safety engineering management in other workplaces:

- Department of Energy (DOE)
- Defense Nuclear Facilities Safety Board (DNFSB)
- Environmental Protection Agency (EPA)
- Department of Transportation (DOT)
- Occupational Safety and Health Administration (OSHA) under Department of Labor (DOL)
- Mine Safety and Health Administration (MSHA) under Department of Labor (DOL)
- Department of Defense (DOD)
- Nuclear Regulatory Commission (NRC)

Among the above, OSHA's mission to prevent work-related injuries, illnesses, and deaths is the driving force behind safety engineering and management for workplaces in the United States.

State and Local Regulations

Section 18 of the Occupational Safety and Health Act of 1970 encourages states to develop and operate their own job safety and health programs to establish safe work environments (OSHA 1970). However, states operating their own state safety and health programs under plans approved by the DOL cover most private-sector workers and are also required to extend their coverage to public-sector (state and local government) workers in the state.

OSHA approves and monitors the state plans. Currently 27 states and other jurisdictions operate approved state plans that cover both private-sector and state and local government employees. Five jurisdictions—Connecticut, Illinois, New Jersey, New York, and the Virgin Islands—cover public employees only.

In addition, the AHJ may use other standards, such as consensus standards (discussed below), as mandatory requirements. For example, many areas have adopted the ASME *Boiler and Pressure Vessel Code* as the mandatory requirement for boiler operations.

Consensus Standards

A *standard* is a practice or a product that is widely recognized or employed because it has been proven best by repeated or common use. A *consensus standard*, similarly, is a practice, procedure, technique, or product that has been widely accepted or applied in a particular area. Consensus standards are developed by technical or professional societies or by national or international standards-setting organizations for consensus agreement among representatives of various interested or affected individuals, companies, organizations, and countries.

Standards developed or adopted by voluntary consensus standards bodies, both domestic and international, are also a basis for safety engineering management. When they are adopted by an AHJ, they become mandatory. Some of these standards are also published in the *Federal Register* and the CFR through a process called *incorporation by reference*.

Incorporation by reference allows federal agencies to comply with the requirement to publish rules in the *Federal Register* by referring to materials already published elsewhere. The legal effect of incorporation by reference is that the material is treated as if it had been published in the *Federal Register* and the CFR. Incorporation by reference is used primarily to make privately developed technical standards federally enforceable. Only the mandatory provisions (i.e., provisions containing the word "shall" or other mandatory language) of standards incorporated by reference are adopted as standards under the Occupational Safety and Health Act (OSHA 1970).

Some commonly used consensus standards are those developed by the American National Standards Institute (ANSI), the National Fire Protection Association (NFPA), the Compressed Gas Association (CGA), the American Society of Mechanical Engineers (ASME), the Institute of Electrical and Electronics Engineers (IEEE), and the Underwriters Laboratory (UL).

International Organizations' Standards

The International Organization for Standardization (ISO) is an international standards-setting body that was founded on February 23, 1947. (*Note:* To eliminate different abbreviations due to the translation of "International Organization for Standardization" into various languages, the short form "ISO" is used universally.) It is a network of the national standards institutes of 159 countries, one member per country, with a central secretariat in Geneva, Switzerland, that coordinates the system. It provides worldwide industrial and commercial standards called ISO standards (ISO 2011).

ISO is a nongovernmental organization. Its members are not delegates of national governments. Many of its member institutes, however, are part of the governmental structure of their countries or are mandated by their governments. Other members have their roots in the private sector. Adoption of ISO standards is voluntary. As a nongovernmental organization, ISO has no legal authority to enforce their implementation, but companies adopting ISO standards help to make international standardization of products or services possible. Companies that adhere to ISO standards assure the quality and reliability of their products and services and also the safety of people and the environment. The most widely accepted standards are ISO 9000 and ISO 14000. ISO 9000 deals with quality management standards (ISO 2005) and ISO 14000 with environmental management standards (ISO 2004). The president of ISO, usually a recognized industry figure, is elected to a two-year term. The secretary general of ISO manages the operations. An ISO council forms the members that develop the proposals. ISO members make the decisions in an annual General Assembly. Detailed information about ISO may be obtained online at www.iso.org.

Other international organizations and foreign governments have rules and regulations that address environmental, safety, and health issues. Some are:

1. International Atomic Energy Agency (IAEA)
2. International Radiation Related Agencies
3. International Environmental Programs Information & Compliance Services
4. The Global Network for Environment & Technology

Best Management Practices

Best management practices are practices or techniques or a combination of practices, procedures, or controls that are *not* required by law or mandatory requirement but have proven to be useful to companies or organizations. Management implements best practices to provide a safe work environment or to safeguard people, property, and the environment. Adopting these policies, procedures, or techniques is considered as adopting best management practices.

The primary elements of the above-mentioned requirements and practices can be summarized in four groups:

1. OSHA program-management regulations
2. OSHA cooperative programs
3. ISO standards
4. EPA program-management regulations

OSHA Program-Management Regulations

Congress created the Occupational Safety and Health Act of 1970 (OSH Act) to prevent work-related injuries, illnesses, and deaths in workplaces and to assign employers the responsibility of providing a safe work environment for their employees. The act applies to employers and their employees in all states, the District of Columbia, Puerto Rico, and all other territories under the federal government. Self-employed persons, family farms that employ family members, workplaces owned and regulated by federal agencies operated under other federal statutes, and state and local governments with their own health and safety programs are not covered by this act. President Richard M. Nixon signed the OSH Act on December 29, 1970, and it became effective in April 1971. It was coauthored by Senator Harrison A. Williams (Dem.-NJ) and Congressman William Steiger (Rep.-WI), so it is also called the Williams-Steiger Act (OSHA 1970).

Before its enactment in the United States, safety and health laws had primarily been left up to the states. Section 5 of the OSH Act mandates the duties of employers and employees and thus sets the main goal of safety engineering management as providing a safe working environment. One of the several functions of the Department of Labor is to improve the work environment of all workplaces in the Unites States. OSHA, which is under the Department of Labor, fulfills that function by the power of the OSH Act of 1970. OSHA sets and enforces standards by providing education, consultation, and partnership for continual improvement in workplaces. OSHA is directed by an assistant secretary of labor who reports to the secretary of labor. The act assigns OSHA two main functions: (1) setting workplace standards, and (2) conducting workplace inspections to ensure that employers provide employees with a safe and healthful workplace and that they comply with occupational safety and health standards. OSHA standards are published in the CFR, Title 29. Federal occupational safety and health standards cover general industry, construction, maritime operations (shipyards, marine terminals, longshoring), and agriculture.

OSHA maintains 29 CFR 1910, Occupational Safety and Health Standards for General Industry, and 29 CFR 1926, Safety and Health Regulations for Construction, and enforces them to ensure the safety and health of the workforces in all workplaces except those that are in states with approved state plans.

A *state plan* is an occupational safety and health program developed by a state that has received the ultimate accreditation from OSHA. The job safety and health standards in state programs have to be "at least as effective as" federal standards. Section 18 of the OSH Act of 1970 encourages states to develop their own programs. Details regarding the process for creating a state plan are available online at www.osha.gov.

All federal agencies of the executive branch of the U.S. government conform to Executive Order 12196 and 29 CFR 1960, Basic Program Elements for Federal Employees OSHA. Section 19 of the act has special provisions for federal employees' safe working conditions. Executive Order 12196 prescribes additional responsibilities for the heads of agencies, the secretary, and the general services administrator (Federal Register 1980). As a requirement, the secretary of labor issues basic program elements for the heads of the agencies according to which they must operate their safety and health programs. These are called Basic Program Elements for Federal Employees OSHA and are available in Title 29, Chapter XVII, Part 1960.

Under the Department of Labor, the Mine Safety and Health Administration (MSHA) ensures mine safety for coal and other mines. Every operator of such a mine and every miner is required to follow the provisions of the Federal Mine Safety and Health Act of 1977. MSHA enforces the mandatory safety and health standards that eliminate accidents and reduce the frequency and severity of accidents or near-misses (MSHA 1977).

Under the heading of OSHA program-management regulations, AHJs, like companies or federal agencies, can include all or some of the federal agency orders, best management practices, and consensus standards such as those created by ASME, ANSI, NFPA, and CGA in addition to the OSHA regulations mentioned above as mandatory requirements in their safety engineering and management policies.

Cooperative Programs

Section 2(b)(1) of the OSH Act encourages employers and employees to reduce the number of occupational safety and health hazards at their places of employment. It also stimulates employers and employees to institute new programs and to perfect existing programs that provide safe and healthful working conditions. Following this section, OSHA created cooperative programs through which businesses and organizations can work with the agency to improve safety and health in workplaces. These initiatives include alliance programs, on-site consultations, Safety and Health Achievement Recognition Programs (SHARP), Voluntary Protection Programs (VPP), and strategic partnerships (OSHA 2011a).

Alliance Program

Alliance programs provide professional and labor organizations, employers, and other groups with tools that help them work together to build relationships with the agency, utilize resources, establish a network with other organizations with the same goal, and obtain recognition for using proactive approaches toward safety and health.

Through alliance programs, OSHA offers help in the areas of training, education, and communication as well as sharing of safety- and health-related information. It helps with the formation of forums and groups to improve workplace safety and health. It addresses occupational trends, emerging issues, and the agency's priorities. In doing so, it emphasizes OSHA's strategic areas to the public by working with businesses, trade and professional organizations, and groups that are involved with safety and health. In an alliance program, OSHA and the participating organization sign a formal document with goals. This program does not relieve the participant organization from OSHA programmed inspections (OSHA 2011b).

On-site Consultation Program

OSHA's On-site Consultation Program primarily helps small businesses, but other businesses may use this program as well. The program consists of free consultation to help identify workplace hazards and improve safety and health. By participating in this program, businesses may even qualify for a one-year exemption from routine OSHA inspections. Most importantly, OSHA will not issue any citation or penalty if hazards are identified and addressed as part of this program. Trained professionals will help to identify hazards and suggest mitigation on site, but the name and any other information provided by the organization is kept confidential. The only commitment requested is for organizations to correct the deficiencies in a timely manner (OSHA 2011c).

Safety and Health Achievement Recognition Program (SHARP)

The Safety and Health Achievement Recognition Program (SHARP) provides useful incentives to employers who have used OSHA's On-site Consultation Program. Employers may seek recognition for their safety and health programs and become models among their peers. SHARP allows workplaces to be exempt from OSHA inspections during the period of SHARP certification. To get into the SHARP program, first an on-site OSHA consultation request has to be made. During this consultation, all hazards are identified and mitigated, as suggested by OSHA professionals. A safety and health program involving employees must be implemented to address at least OSHA's 1989 Safety and Health Program Management Guidelines. Also, the days away, restricted, or transferred (DART) rate and total recordable case (TRC) rate must be below the national average, and the company has to agree to notify the state Consultation Project Office before making any change in working conditions or introducing a new hazard (OSHA 2011c).

OSHA Strategic Partnership Program (OSPP)

The OSHA Strategic Partnership Program (OSPP) partners companies with OSHA to address specific safety and health issues. This partnership can be with one or more organizations, employees, and employee representatives. Instead of the usual role as the enforcer of safety and health standards, here OSHA serves as a technical resource and facilitator to employers, employees, unions, trade associations, state on-site consultation projects, and other interested parties. It helps them to use its resources to train employees and develop site-specific safety and health management systems. The OSHA Strategic Partnership (OSPP) is also a voluntary program. Each partnership prepares its own formal agreement with specific goals, strategies, and performance measures. This program is available to private as well as government agencies (OSHA 2011d).

Voluntary Protection Program (VPP)

The Voluntary Protection Program (VPP) is the most important program in this category. This program promotes safety and health excellence through cooperative efforts among employees, management, unions, and OSHA. VPP has performance-based criteria, and any safety and health system of a workplace may qualify for one of three programs: Star, Merit, or Star Demonstration. Interested sites must apply. OSHA reviews the application and then conducts a thorough on-site evaluation of the safety and health program and its implementation. As a VPP-certified site, the establishment will not receive OSHA compliance inspections unless it fails to maintain its VPP status or other significant safety and health issues arise.

Along with this benefit, statistical evidence shows fewer injuries and illnesses for organizations that have implemented the program. The success is impressive: reductions in injuries and illnesses begin when the site commits to the VPP approach to safety and health management and in the middle of application process.

Fewer injuries and illnesses also mean greater profits as workers' compensation premiums and other costs are reduced. The reductions in injuries and illnesses are achieved by the principles of VPP: management leadership, employee involvement, work-site analysis, hazard prevention and control, and safety and health training (OSHA 2011a).

OSHA Challenge Program

Any employee, company, or government agency can take a proactive approach toward going through the OSHA Challenge Program to prepare for the Voluntary Protection Program's Star, Merit, or Star Demonstration status. The OSHA Challenge Program is applicable for general industry as well as construction work activities. There are three stages that a participant has to go through in this program, and an online road map guides the company to improve its safety and health program.

Each participant company has to be associated with an administrator of this program. Some companies may qualify to be an administrator for their own facilities, but administrators may not be private safety or health consultants or for-profit associations. The OSHA Challenge Program will be in operation for at least two years; at that time, OSHA will evaluate the participant's safety and health program. Based on that evaluation, the participant will graduate, continue, or terminate the program. The program benefits a participant company by substantially improving its safety and health program (OSHA 2011e).

International Organization for Standardization (ISO) Standards

ISO produces voluntary standards that are considered useful to industrial and business organizations of all types, to governments and other regulatory bodies, to the suppliers and customers of products and services in both public and private sectors, and, ultimately, to people in general in their roles as consumers and end users. ISO standards contribute to making the development, manufacturing, and supplying of products and services more efficient, safer, and cleaner. They make trade between countries easier and fairer. They provide governments with a technical base for health, safety, and environmental legislation, which is why they are used in safety engineering and management along with other requirements. ISO 9000 and ISO 14000 are international references for quality management and environmental management, respectively. Industries and businesses in many countries are making these standards mandatory, along with regulations and laws in the safety engineering and management arena, to achieve consistency in quality, safety, and environmental areas.

EPA Program-Management Regulations

More than a dozen major statutes or laws form the legal basis for the programs of the Environmental Protection Agency (EPA).

National Environmental Policy Act of 1969 (NEPA)

The purpose of this act is to prevent or eliminate damage to the environment and biosphere by considering the impact of any proposed action and reasonable alternatives. The act requires federal agencies to include the understanding of ecological systems and natural resources in their decision-making process and to establish the Council on Environmental Quality (CEQ). Federal agencies must prepare detailed reports, known as environmental impact statements (EISs). The Environmental Protection Agency (EPA) confirms that it complies with NEPA and reviews and provides comments on EISs prepared by other federal agencies. EPA also maintains a national filing system for all EISs. Title II of NEPA asks the Council on Environmental Quality (CEQ) to gather environmental quality information, evaluate federal programs according to Title I of NEPA, develop and promote national policies, and conduct studies and research on ecosystems and environmental quality (EPA 2011a).

The Clean Air Act of 1970 (CAA), Amended in 1990

The Clean Air Act is a comprehensive federal law that regulates air emissions from area, stationary, and mobile sources. This law authorizes the EPA to establish National Ambient Air Quality Standards (NAAQS) to protect public health and the environment. The act prevents significant deterioration of air quality in the country by setting limits on how much of a pollutant can be in the air anywhere in the United States. States must develop state implementation plans (SIPs), which must be at least as stringent as the CAA, and the EPA must approve each SIP after review. The 1990 CAA amendments provide control over interstate air pollution, international air pollution that originates in Mexico and Canada and drifts into the United States, and pollution that travels from the United States to Mexico and Canada. It also has a permit program and gives authority to the EPA for enforcement of the law. The details of the act are in 40 CFR Parts 50–99 (EPA 2011b).

Federal Insecticide, Fungicide and Rodenticide Act of 1972 (FIFRA)

The Federal Insecticide, Fungicide and Rodenticide Act (FIFRA) provides federal control of the distribution, sale, and use of pesticides. All pesticides used in the entire country must be registered by the EPA. They must be properly labeled and used according to the specifications. Under FIFRA, the EPA was given authority to study the consequences of pesticide usage and to require users (farmers, utility companies, and others) to register when purchasing pesticides to protect applicators and consumers of pesticides and insecticides and also to protect the environment. Users, if applying the regulated material, also must take examinations for certification as pesticide applicators. The details of the act are available in 40 CFR Parts 150–189 (EPA 2011c).

The Endangered Species Act of 1973 (ESA)

The Endangered Species Act (ESA) provides a program for the conservation of threatened and endangered plants and animals and the habitats in which they are found by conserving their ecosystems. The U.S. Fish and Wildlife Service of the Department of Interior maintains a list of 632 endangered species and 190 threatened species. The details are available in 50 CFR Part 17 (EPA 2011d).

The Safe Drinking Water Act of 1974 (SDWA), Amended in 1986 and 1996

The Safe Drinking Water Act (SDWA) authorizes the EPA to establish safe standards of purity for all water actually or potentially designed for drinking use from aboveground and underground sources. The SDWA authorizes EPA to create combined federal, state, and tribal systems to comply with the standard. It requires all owners or operators of public water systems to comply with primary (health-related) standards. State governments, which assume this power from the EPA, also encourage the attainment of secondary standards that are nuisance-related. The amendments in 1986 and 1996 extended the law to cover protection at the source through the control of underground injection of liquid waste, operator training, and information for the public.

The EPA developed two water-quality standards—primary and secondary. The primary standard is legally enforceable and applies to public water systems. It protects drinking water quality by limiting the levels of specific contaminants that can adversely affect public health and are known or anticipated to occur in water. The primary standard lists a maximum containment level (MCL)—the amount of a contaminant allowed in water delivered to a user of any public water system—or a treatment technique (TT), a procedure or level of technological performance set when there is no reliable method to measure a contaminant at very low levels.

Secondary standards are nonenforceable guidelines regarding contaminants that may cause cosmetic effects (such as skin or tooth discoloration) or aesthetic effects (such as changes in taste, odor, or color) in drinking water. The EPA recommends secondary standards to water systems but does not require systems to comply. However, states may choose to adopt them as enforceable standards.

Secondary standards are based on nonenforceable maximum containment-level goals (MCLGs). The MCLG of a drinking water contaminant is the level below which there is no known or expected health risk. MCLGs allow for a margin of safety. They are based on the risks of exposure to infants, the elderly, and persons with compromised immune systems. Drinking water standards apply to public water systems (PWSs) that provide water for human consumption through at least fifteen service connections or that regularly serve at least 25 individuals. Public water systems include municipal water companies, homeowner associations, businesses, campgrounds, schools, and shopping malls.

The details of this act are available in 40 CFR Parts 141–149 (EPA 2011e).

The Resource Conservation and Recovery Act of 1976 (RCRA)

The Resource Conservation and Recovery Act (RCRA), also known as the "cradle to grave" rule, gives the EPA the authority to control hazardous waste from its generation through transportation, treatment, storage, and ultimately disposal. The RCRA also sets forth a framework for the management of nonhazardous wastes. The goal of this act is to protect people and the environment from the harmful effects of disposed waste; to clean up leaked, spilled, or improperly stored waste; and to advocate reuse, reduction, and recycling.

The 1986 amendments to the RCRA enable the EPA to address environmental problems that could result from underground tanks storing petroleum and other hazardous substances. The RCRA focuses only on active and future facilities and does not address abandoned or historical sites. The details of this act are available in 40 CFR Parts 261–299 (EPA 2011f).

The Toxic Substances Control Act of 1976 (TSCA)

The Toxic Substances Control Act (TSCA) gives the EPA broad authority to regulate the manufacture, use, distribution in commerce, and disposal of chemical substances. It requires the EPA to review the health and environmental effects of existing chemical substances and all new chemicals before they are manufactured for commercial purposes and to control some substances that have been identified as potentially high risks to the public. The TSCA is a federally managed law and is not delegated to states. It is overseen by the EPA Office of Pollution Prevention and Toxics (OPPT). The TSCA became law on October 11, 1976, but Congress later added more titles. The original part remained as Title I, Control of Hazardous Substances and Asbestos; Hazard Emergency Response was added as Title II; Indoor Air Radon Abatement was added as Title III; and Lead-Based Paint Exposure became Title IV (EPA 2011g).

The Clean Water Act of 1977 (CWA)

This is officially the federal Water Pollution Control Act, but it is commonly known as the Clean Water Act (CWA). It regulates both direct and indirect discharges of water. The goal of this act is to "restore and maintain the chemical, physical, and biological integrity of the nation's waters by preventing point and nonpoint sources, providing assistance to publicly owned treatment works (POTWs) for the improvement of wastewater treatment, and maintaining the integrity of wetlands." The act employs a variety of regulatory and nonregulatory tools to reduce direct pollutant discharges into waterways, finance municipal wastewater treatment facilities, and manage polluted runoff. It does not deal directly with groundwater or with water quantity issues. The details of this act are available in 40 CFR, Parts 100–149 (EPA 2011h).

Comprehensive Environmental Response, Compensation, and Liability Act of 1980 (CERCLA)

The Comprehensive Environmental Response, Compensation, and Liability Act (CERCLA), commonly known as Superfund,

1. establishes prohibitions and requirements concerning closed and abandoned hazardous waste sites
2. provides for liability of persons responsible for releases of hazardous waste at these sites

3. establishes a trust fund to provide for clean up when no responsible party can be identified.

The law authorizes the EPA to enforce clean up by the responsible parties or to force responsible parties to reimburse the Superfund for clean up. The EPA implements the act in all 50 states and U.S. territories. State environmental protection or waste management agencies coordinate all Superfund site identification, monitoring, and response activities (EPA 2011i).

The Emergency Planning & Community Right to Know Act of 1986 (EPCRA)

After two accidents related to chemicals, one in Bhopal, India, (1984) and the other in Institute, West Virginia, (1985), Congress enacted the Emergency Planning & Community Right to Know Act (EPCRA) of 1986 to protect communities. It contains requirements regarding emergency planning programs, emergency release notification, community right-to-know reporting, and toxic chemical release reporting.

EPCRA requires each state to appoint a state emergency response commission (SERC). The SERCs are required to divide their states into emergency planning districts and to name a local emergency planning committee (LEPC) for each district. To ensure that all necessary elements of the planning process are covered, committee members include health officials, government and media representatives, community groups, industrial facilities, fire fighters, and emergency managers. This law helps local communities protect public health, safety, and the environment from chemical hazards. It is also known as SARA, Title III, and is commonly referred to as the Community Right to Know law. Details of the act are available in 40 CFR, Part 355 (EPA 2011j).

Federal Food, Drug, and Cosmetic Act (FFDCA)

FFDCA is the basic authority intended to ensure that:

- foods are safe to eat and produced under sanitary conditions
- drugs and devices are safe and effective for their intended uses
- cosmetics are safe and made from appropriate ingredients
- all labeling and packaging is truthful, informative, and not deceptive

The Food and Drug Administration (FDA) is primarily responsible for enforcing the FFDCA, although the USDA also has some enforcement responsibility. The EPA establishes limits for concentrations of pesticide residues on food under this act. This act is known for naming certified food color additives like Brilliant Blue FCF (FD&C Blue No. 1), Erythrosine (FD&C Red No. 3), and so on (EPA 2011k).

The Superfund Amendments and Reauthorization Act of 1986 (SARA)

The Superfund Amendments and Reauthorization Act (SARA) amended the Comprehensive Environmental Response, Compensation, and Liability Act (CERCLA) on October 17, 1986.

This act has several key features:

- It increases the size of the Superfund.
- It expands the response authority of the EPA.
- It strengthens enforcement activities at Superfund sites.
- It broadens the law to include federal facilities.
- It adds a citizen suit provision.
- It allows the EPA to condemn property.
- It provides deadlines for response action.

Also, SARA requires the EPA to revise the Hazard Ranking System (HRS) to ensure that it accurately assesses the relative degree of risk to human health and to the environment posed by hazardous waste sites and disposal facilities.

In addition, SARA authorizes states to participate in clean-up processes, from initial site assessment to selecting and carrying out the remedial action and negotiating with responsible parties. To encourage states to establish new hazardous waste treatment and disposal facilities, SARA requires that states assure that they will have adequate disposal capacity for all hazardous wastes expected to be generated within the state for the next twenty years. This requirement went into effect in November 1989 (EPA 2011l).

The Pollution Prevention Act of 1990 (PPA)

The Pollution Prevention Act (PPA) requires that:

- pollution should be prevented or reduced at the source whenever feasible
- pollution that cannot be prevented should be recycled in an environmentally safe manner whenever feasible
- pollution that cannot be prevented or recycled should be treated in an environmentally safe manner whenever feasible
- disposal or other release of pollution into the environment should be used only as a last resort and conducted in an environmentally safe manner (EPA 2011m)

The Oil Pollution Act of 1990 (OPA)

The Oil Pollution Act became a law in August 1990. It provides the EPA the ability to prevent and respond to catastrophic oil spills. A trust fund financed by a tax on oil is available to clean up spills when the responsible party is incapable of doing so or unwilling to do so. The federal government directs response efforts for certain types of spills. Area committees are formed of federal, state, and local government officials, and the committees develop site-specific area contingency plans. Certain oil storage facilities and vessels that may cause serious damage to the environment are required to submit their own plans to the federal government giving details of how they will respond to large discharges. Details are available in 33 U.S.C. 2701–2761 (EPA 2011n).

Chemical Safety Information, Site Security and Fuels Regulatory Relief Act (CSISSFRRA), January 6, 1999

This is an amendment to Section 112(r) of the Clean Air Act.

Under section 112(r), facilities handling large quantities of extremely hazardous chemicals are required to include information about handling the chemicals in a risk management plan (RMP) submitted to the EPA. As required by CSISSFRRA, this provides members of the public and government officials with access to that information in ways designed to minimize the likelihood of accidental releases, the risk to national security associated with posting the information on the Internet, and the likelihood of harm to public health and welfare (EPA 2011o).

There are several other laws in the environmental safety area, such as the Lead-Based Paint Poisoning Prevention Act, 1971; the Hazardous Materials Transportation Act, 1975; and the Asbestos Hazard Emergency Response Act, 1986.

CONCLUSION

The regulatory requirements applicable to the management of safety engineering work depend on the scope of the work, identification of the hazards associated with the scope of the work, the applicable hazard control and mitigation process, and the location and responsibility/ownership of the process, project, or facility. Location and ownership will be determining factors with regard to which rules, regulations, codes, or standards are applicable for that process, project, or facility. Depending on the level of acceptable risk and the level of protection set by the management, the above-mentioned requirements may be incorporated into the appropriate policies and procedures of a company or agency using a process that balances risk, cost, and availability of technology.

In the author's opinion, for the management of any safety engineering work, six elements should be considered:

1. development of strategy and policy
2. implementation of procedures, including roles and responsibilities
3. identification of controls and requirements according to the scope of work
4. training
5. audits and inspections
6. continual improvement based on feedback and lessons learned from audits and inspections

The following are steps organizations may consider adopting in order to establish necessary controls and requirements for managing safety engineering work:

1. Identify the total scope of work. This will provide information about the laws and regulations that must be applied. For example, if the scope of work includes environmental or waste management work, identify which environmental laws and regulations are applicable.
2. Identify the location of the work. This will define whether it falls under federal agency, private industry, state, or local authority.
3. Identify the authority having jurisdiction (AHJ). This is over and above what is identified in the previous step. For example, an electrical safety committee set up by any one company's management is the AHJ that will decide whether the company will adopt NFPA 70E *Standard for Electrical Safety in the Workplace* as mandatory.
4. Ascertain whether federal OSHA or a state plan will apply. This will be based on the ownership of the workplace.
5. Determine whether work falls under operations and maintenance or construction and then decide whether 29 CFR 1910, General Industry, or 29 CFR 1926, Construction, or both, apply. This decision is based on the type and scope of work.
6. Decide which environmental law shall apply. This is based on the work involved.
7. Ascertain the acceptable risk level using probability and consequence as the basis of risk assessment. This will depend on management policy.
8. Identify which consensus standards and best management practices should be adopted. This depends on management policy.
9. Review whether any specific federal agency orders, such as DOT orders, are applicable.
10. Determine whether ISO standards must be applied.

References

Centers for Disease Control (CDC) (last accessed on November 21, 2006). www.cdc.gov/niosh/homepage.html

Environmental Protection Agency (EPA). 2000. Executive Order 13148, *Greening the Government Through Leadership in Environmental Management* (accessed October 18, 2011). www.epa.gov/epa/pubs/eo13148.pdf

_____. 2011a. *National Environmental Policy Act* (accessed October 21, 2011). www.epa.gov/compliance/nepa

_____. 2011b. *Summary of the Clean Air Act* (Accessed October 21, 2011). www.epa.gov/lawsregs/laws/caa.html

_____. 2011c. *Summary of the Federal Insecticide, Fungicide, and Rodenticide Act* (accessed October 21. 2011). www.epa.gov/lawsregs/laws/fifra.html

_____. 2011d. *Summary of the Endangered Species Act* (accessed October 21, 2011). www.epa.gov/lawsregs/laws/esa.html

_____. 2011e. *Summary of the Safe Drinking Water Act* (accessed October 21, 2011). www.epa.gov/lawsregs/laws/sdwa.html

_____. 2011f. *Summary of the Resource Conservation and Recovery Act* (accessed October 21. 2011). www.epa.gov/lawsregs/laws/rcra.html

_____. 2011g. *Summary of the Toxic Substances Act* (accessed October 21, 2011). www.epa.gov/lawsregs/laws/tsca.html

_____. 2011h. *Summary of the Clean Water Act* (accessed October 21, 2011). www.epa.gov/lawsregs/laws/cwa.html

_____. 2011i. *Summary of the Comprehensive Environmental Response, Compensation and Liability Act* (accessed October 21.2011). www.epa.gov/lawsregs/laws/cercla.html

_____. 2011j. *Summary of the Emergency Planning and Community Right-to-Know Act* (accessed October 21, 2011). www.epa.gov/lawsregs/laws/epcra.html

_____. 2001k. *Summary of the Federal Food, Drug and Cosmetic Act* (accessed October 21, 2011). www.epa.gov/lawsregs/laws/ffdca.html

_____. 2011l. *SARA Overview* (accessed October 21, 2011). www.epa.gov/superfund/policy/sara.html

_____. 2011m. *Summary of the Pollution Prevention Act* (accessed October 21, 2011). www.epa.gov/lawsregs/laws/ppa.html

_____. 2011n. *Summary of the Oil Pollution Act* (accessed October 21. 2011). www.epa.gov/lawsregs/laws/opa.html

_____. 2011o. *Summary of the Chemical Safety Information, Site Security, and Fuels Regulatory Relief Act* (accessed October 21, 2011). www.epa.gov/lawsregs/laws/csissfrra.html

Federal Register. 1980. Executive Order 12196, *Occupational Safety and Health Programs for Federal Employees* (accessed October 18, 2011). www.archives.gov/federal-register/codification/executive-order/12196.htm

International Organization for Standardization (ISO). 2004. ISO 14001-2004, *Environmental Management—Environ-*

mental Communication—Guidelines and Examples (accessed October 17, 2011). www.iso.org/iso/iso_14000_essentials

———. 2005. ISO 9000-2005, *Quality Management Systems--Fundamentals and Vocabulary* (accessed October 17, 2011). www.iso.org/iso/iso_9000_essentials

———. 2011. *About ISO* (accessed October 18, 2011). www.iso.org

Mine Safety and Health Administration (MSHA). *MSHA Statutory Functions* (accessed October 21, 2011). www.msha.gov/MSHAINFO/MSHAINF1.HTM

Occupational Safety and Health Administration (OSHA). 1970. *Occupational Safety and Health Act* (accessed October 17, 2011). www.osha.gov/pls/oshaweb/owasrch.search_form?p_doc_type=oshact

———. 2011a. *The OSHA Alliance Program* (accessed October 17, 2011). www.osha.gov/dcsp/alliances/whatis.html

———. 2011b *OSHA Challenge: A Roadmap to Safety and Excellence* (accessed October 17, 2011). www.osha.gov/dcsp/vpp/challenge.html

———. 2011c. *On-Site Consultation Program: Safety and Health Achievement Recognition Program (SHARP)* (accessed October 17, 2011). www.osha.gov/dcsp/sharp.index.html

———. 2011d. *OSHA Strategic Partnership Program* (accessed October 17, 2011). www.osha.gov/dcsp/partnerships.index.html

———. 2011e. *Voluntary Protection Program* (accessed October 17, 2011). www.osha.gov/dcsp/vpp/index.html

APPENDIX: ADDITIONAL RESOURCES

Chemical Safety Information, Site Security and Fuels Regulatory Relief Act of 1999 (CSISSFRRA), 42 U.S.C. Section 7410 et seq. (accessed October 20, 2011). www.dotcr.ost.dot.gov/documents/ycr/PL106-40.pdf

Clean Air Act of 1990 (CAA), 42 U.S.C. Section 7401 et seq. (accessed October 18, 2011). www.epa.gov/air/caa

Clean Water Act of 1977 (CWA), 33 U.S.C. Section 1251 et seq. (accessed October 20, 2011). epa.senate.gov/water.pdf

Comprehensive Environmental Response, Compensation and Liability Act of 1980 (CERCLA), 42 U.S.C. Section 9601 et seq. (accessed October 20, 2011). epa.senate.gov/cercla.pdf

Emergency Planning and Community Right-to-Know Act of 1986 (EPCRA), 42 U.S.C. Section 11001 et seq. (accessed October 18, 2011). frwebgate.access.gpo.gov/cgibin/usc.cgi?ACTION=BROWSE&TITLE=42USCC116

Endangered Species Act of 1973 (ESA), 16 U.S.S. 1531 et seq. (accessed October 18, 2011). epa.senate.gov/esa1973.pdf

Federal Food, Drug and Cosmetic Act (FFDCA), 21 U.S.C. 301 et seq. (accessed October 20, 2011). epa.senate.gov/FDA_001.pdf

Federal Insecticide, Fungicide and Rodenticide Act of 1972 (FIFRA), 7 U.S.C. 136 et seq. (accessed October 20, 2011). www.epa.gov/opp00001/regulating/fifra/pdf

Mine Safety and Health Administration (MSHA). *Federal Mine Safety and Health Act of 1977* (accessed October 17, 2011). www.msha.gov/regs/act/acttc.html

National Environmental Policy Act of 1969 (NEPA) (accessed October 17, 2011). ceq.hss.doe.gov/laws=and_exec_orders/the_nepa_statute.html

Oil Pollution Act of 1990 (OPA), 33 U.S.C. Section 2701 et seq. (accessed October 20, 2011). epa.senate.gov/opa90.pdf

Pollution Prevention Act of 1990 (PPA), 42 U.S.C. Section 13101 et seq. (accessed October 21, 2011). epa.senate.gov/PPA90.pdf

Resource Conservation and Recovery Act of 1976 (RCRA), 42 U.S.C. Section 6901 et seq. (accessed October 20, 2011). epa.senate.gov/rcra.pdf

Safe Drinking Water Act of 1974 (SDWA), 42 U.S.C. Section 300f et seq. (accessed October 21, 2011). epa.senate.gov/sdwa.pdf

Toxic Substances Control Act of 1976 (TSCA), 15 U.S.C. Section 2601 et seq. (accessed October 21, 2011). epa.senate.gov/tsca.pdf

General Safety Management

2

Jeffery C. Camplin

LEARNING OBJECTIVES

- Recognize relevant concepts and components of an effective safety management system.

- Express voluntary safety management concept models developed by OSHA and ANSI for effective safety management.

- Analyze safety management system needs and integrate safety management into the business culture of an organization.

- Express the relevant concepts of an effective safety professional including education, experience, and certification.

DEFINING AN EFFECTIVE safety management system first requires a definition of the function of safety. Defining safety is easy. Petersen states, "The function of safety is to locate and define the operational errors that allow accidents to occur. This function can be carried out in two ways: (1) by asking why accidents happen—searching for their root causes—and (2) by asking whether certain known effective controls are being utilized" (Petersen 2003).

Although the specific job descriptions and titles of those delegated with the safety function may vary among organizations, the goal of safety professionals is to safeguard the entities' assets. First and foremost, they protect the human assets; second, they manage the tangible and intangible assets in a cost-effective manner (Schneid 2000).

Defining a safety management system can be more difficult. Most safety management systems are unique, differing in structure, definition, and/or implementation (Colvin 1992). Addressing safety and health issues in the workplace saves the employer money and adds value to the business. Recent estimates place the business costs associated with occupational injuries at close to $170 billion—expenditures that come straight out of company profits (OSHA 2005). When workers stay whole and healthy, OSHA found that the direct cost savings to businesses can include:

- lower workers' compensation insurance costs
- reduced medical expenditures
- smaller expenditures for return-to-work programs
- fewer faulty products
- lower costs for job accommodations for injured workers
- less money spent for overtime benefits

Safety and health can also make big reductions in indirect costs, due to:

- increased productivity
- higher-quality products
- increased morale
- better labor/management relations
- reduced turnover
- better use of human resources

Employees and their families benefit from safety and health because

- their incomes are protected
- their family lives are not hindered by injury
- their stress is not increased

Most successful safety management systems share common elements. This chapter will discuss: (1) the common elements of any safety management system; (2) tools of the trade to develop and implement these system elements; and (3) an overview of competencies and resources for improving the safety, health, and environmental (SHE) professional.

Defining an Effective Safety and Health Management System

What is a safety and health program or management system? There are many ways to answer this question. Below are some definitions.

> An effective safety and health program depends on the credibility of management's involvement in the program, inclusion of employees in safety and health decisions, rigorous worksite analysis to identify hazards and potential hazards, including those which could result from a change in worksite conditions or practices, stringent prevention and control measures, and thorough training. It addresses hazards whether or not they are regulated by government standards. (OSHA 1998a, 3905)

> Measures for the prevention and control of occupational hazards in the workplace should be based upon a clear, implementable and well-defined policy at the level of the enterprise. The occupational health and safety policy represents the foundation from which occupational health and safety goals and objectives, performance measures, and other system components are developed. It should be concise, easily understood, approved by the highest level of management, and known by all employees in the organizations. (IOHA 1998, B-1)

There are several themes that emerge from effective safety management systems. The first is the involvement by all employees in a company, including top management. There must be buy-in by all employees at all levels of the organization. Ownership of safety management is not the sole charge of the SHE professional. Ownership of a company's safety management system should emanate from the senior management team as part of the overall business plan "inseparable from productivity and profitability" (Barfield 2004, 8). Management must clearly define realistic goals and expectations that hold all levels of the organization accountable for the program's success. The seamless integration of safety management throughout an organization involves assessing its overall business culture. "An organization's culture determines the probability of success of its hazard management endeavors" (Manuele 2003, 1). To improve safety awareness, it is also important to assess how the view of workplace safety by an organization's management might be removed "from those on the shop floor" (Petersen 2004, 29). The SHE professional will need a strategy for providing senior management with the tools and guidance for safety management system success. These tools and guidance include establishing reasonable and achievable goals, objectives, and performance measures that are easily understood throughout all levels of an organization. Attaining management's understanding of safety management requires that the SHE professional talk the language of business. Safety should not be seen as a cost for compliance, but more like a return on investment, in the form of savings to a company's bottom line. The SHE professional's role in providing tools and guidance to an organization's safety management system are discussed further in this chapter.

A second theme found in the definitions of a safety management system involves work-site analysis and hazard recognition. The OSHA definition provided earlier in the chapter discusses a "rigorous worksite analysis to identify hazards and potential hazards, including those which could result from a change in worksite conditions or practices." The SHE professional cannot provide guidance on preventing and controlling workplace hazards unless the hazards are identified and analyzed with an evaluation of the significance of risks derived from them. (Manuele 2005). Hazard identification and analysis involves a variety of worksite examinations to identify not only existing hazards, but also conditions and operations in which changes might create hazards. Effective management actively analyzes the work and the work site to anticipate and prevent harmful occurrences (OSHA 2002). There are several techniques used to analyze workplace hazards and associated risks. Manuele found commonly used techniques include preliminary hazard analysis, safety reviews, operations analysis, what-if analysis, hazard and operability analysis (HAZOP), failure modes and effects analysis, fault tree analysis, and management oversight and risk tree (Manuele 2005).

A third theme in safety management is effective hazard control and prevention. The SHE professional often is in the position of recommending or providing guidance on solving workplace hazards. These recommendations must address specific hazards and associated risks if the intended risk reduction is to be achieved. Figure 1 illustrates the safety decision hierarchy that will help the SHE professional and an organization's management understand how to evaluate and effectively resolve unacceptable hazardous situations (Manuele 2005).

Manuele provides a simple hazard/risk problem-solving methodology: (1) identify and analyze the problem, (2) consider actions in order of effectiveness, (3) decide and take action, and (4) measure for effectiveness and reanalyze as needed. Effective hazard control and prevention requires ensuring that actions selected to solve workplace hazards accomplish their intended purpose of risk reduction. The final theme in safety management is the development of effective training to provide the knowledge and skills necessary for all employees to understand workplace hazards and safe procedures.

Structuring the safety management system can be accomplished in many ways. A management systems approach has been developed by OSHA as a guideline and a consensus standard safety management system has been developed by industry stakeholders. The American National Standards Institute (ANSI) has developed a voluntary consensus standard for occupational safety and health management to help organizations minimize workplace risks and reduce the occurrence and cost of occupational injuries, illnesses, and fatalities. The Z10 standard (ANSI/AIHA 2005) is designed to continually improve safety and health performance and is aligned with the traditional Plan—Do—Check—Act approach for improving the workplace (Walton 1986).

The consensus standard provides basic requirements for occupational health and safety management systems rather than detailed specifications to provide flexibility in a manner appropriate to each organization and corresponding with its occupational health and safety risks. The standard defines *what* has to be accomplished in generic performance terms, but it leaves the *how* to each organization to develop. The standard recognizes that the risks, organizational structure, culture, and other characteristics of each organization are unique, and that each organization has to define its own specific measures of performance. An apparent theme throughout the standard is that hazards are to be identified and evaluated; risks are to be assessed and prioritized; and risk elimination, reduction, or control measures are to be taken to assure that an acceptable risk level is attained.

To properly understand and implement the Z10 standard, the SHE professional should know the ANSI definitions of hazard and risk. A *hazard* is defined as a condition, set of circumstances, or inherent property that can cause injury, illness, or death. *Risk* is defined as

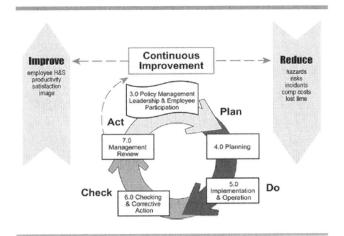

FIGURE 1. Continuous improvement diagram
(*Source:* ANSI/AIHA 2005)

an estimate of the combination of the likelihood of an occurrence of a hazardous event or exposure(s) and the severity of injury or illness that may be caused by the event or exposures (ANSI/AIHA 2005). Additional information on the Z10 standard will be discussed throughout the chapter.

OSHA has also developed a management approach to merge the themes discussed above into a system for building an effective safety management system. This OSHA management system has four elements:

1. Management, leadership, and employee involvement
2. Work-site analysis
3. Hazard prevention
4. Safety and health training

This chapter next looks at each of these elements and other resources available from OSHA. Then some tools of the trade are provided to assist the SHE professional in the implementation of an effective safety management system.

OSHA's Role in Safety Management

Section 5(a)(1) of the OSH Act, often referred to as the General Duty Clause, requires an employer to "furnish to each of his employees employment and a place of employment, which are free from recognized hazards that are causing or are likely to cause death or serious physical harm to his employees." Section 5(a)(2) requires employers to "comply with occupational safety and health standards" promulgated under this Act (OSHA 1970). However, OSHA compliance does not necessarily mean an employer has a complete and effective safety management system. In 1998, OSHA published a draft rule for safety management systems based upon their General Duty clause (see Sidebar 1).

OSHA has recognized that regulatory compliance is only one factor in the implementation of a successful safety system. Over their years of experience with enforcing the provisions of the Occupational Safety and Health Act of 1970 (OSH Act 1970), OSHA representatives have noted a strong correlation between the application of sound management

---

SIDEBAR 1

DRAFT PROPOSED SAFETY AND HEALTH PROGRAM RULE (OSHA 1998a)
29 CFR 1900.1
Docket No. S&H-0027

What is the purpose of this rule? The purpose of this rule is to reduce the number of job-related fatalities, illnesses, and injuries. The rule will accomplish this by requiring employers to establish a workplace safety and health program to ensure compliance with OSHA standards and the General Duty Clause of the Act (Section 5(a)(1)).

(a) Scope.

(a)(1) Who is covered by this rule? All employers covered by the Act, except employers engaged in construction and agriculture, are covered by this rule.

(a)(2) To what hazards does this rule apply? This rule applies to hazards covered by the General Duty Clause and by OSHA standards.

(b) Basic obligation.

(b)(1) What are the employer's basic obligations under the rule? Each employer must set up a safety and health program to manage workplace safety and health to reduce injuries, illnesses, and fatalities by systematically achieving compliance with OSHA standards and the General Duty Clause. The program must be appropriate to conditions in the workplace, such as the hazards to which employees are exposed and the number of employees there.

(b)(2) What core elements must the program have? The program must have the following core elements:

(i) Management leadership and employee participation;
(ii) Hazard identification and assessment;
(iii) Hazard prevention and control;
(iv) Information and training; and
(v) Evaluation of program effectiveness.

(b)(3) Does the rule have a grandfather clause? Yes. Employers who have implemented a safety and health program before the effective date of this rule may continue to implement that program if:

(i) The program satisfies the basic obligation for each core element; and
(ii) The employer can demonstrate the effectiveness of any provision of the employer's program that differs from the other requirements included under the core elements of this rule.

General Safety Management

(c) Management leadership and employee participation.

(c)(1) Management leadership.

(c)(1)(i) What is the employer's basic obligation? The employer must demonstrate management leadership of the safety and health program.

(c)(1)(ii) What must an employer do to demonstrate management leadership of the program? An employer must:

(A) Establish the program responsibilities of managers, supervisors, and employees for safety and health in the workplace and hold them accountable for carrying out those responsibilities;

(B) Provide managers, supervisors, and employees with the authority, access to relevant information, training, and resources they need to carry out their safety and health responsibilities; and

(C) Identify at least one manager, supervisor, or employee to receive and respond to reports about workplace safety and health conditions and, where appropriate, to initiate corrective action.

(c)(2) Employee participation.

(c)(2)(i) What is the employer's basic obligation? The employer must provide employees with opportunities for participation in establishing, implementing, and evaluating the program.

(c)(2)(ii) What must the employer do to ensure that employees have opportunities for participation? The employer must:

(A) Regularly communicate with employees about workplace safety and health matters;

(B) Provide employees with access to information relevant to the program;

(C) Provide ways for employees to become involved in hazard identification and assessment, prioritizing hazards, training, and program evaluation;

(D) Establish a way for employees to report job-related fatalities, injuries, illnesses, incidents, and hazards promptly and to make recommendations about appropriate ways to control those hazards; and

(E) Provide prompt responses to such reports and recommendations.

(c)(2)(iii) What must the employer do to safeguard employee participation in the program? The employer must not discourage employees from making reports and recommendations about fatalities, injuries, illnesses, incidents, or hazards in the workplace, or from otherwise participating in the workplace safety and health program.

Note: In carrying out this paragraph (c)(2), the employer must comply with the National Labor Relations Act.

(d) Hazard identification and assessment.

(d)(1) What is the employer's basic obligation? The employer must systematically identify and assess hazards to which employees are exposed and assess compliance with the General Duty Clause and OSHA standards.

(d)(2) What must the employer do to systematically identify and assess hazards and assess compliance? The employer must:

(i) Conduct inspections of the workplace;

(ii) Review safety and health information;

(iii) Evaluate new equipment, materials, and processes for hazards before they are introduced into the workplace; and

(iv) Assess the severity of identified hazards and rank those that cannot be corrected immediately according to their severity.

Note: Some OSHA standards impose additional, more specific requirements for hazard identification and assessment. This rule does not displace those requirements.

(d)(3) How often must the employer carry out the hazard identification and assessment process? The employer must carry it out:

(i) Initially;

(ii) As often thereafter as necessary to ensure compliance with the General Duty Clause and OSHA standards and at least every two years; and

(iii) When safety and health information or a change in workplace conditions indicates that a new or increased hazard may be present.

(d)(4) When must the employer investigate safety and health events in the workplace? The employer must investigate each work-related death, serious injury or illness, or incident (near-miss) having the potential to cause death or serious physical harm.

(d)(5) What records of safety and health program activities must the employer keep? The employer must keep records of the hazards identified and their assessment and the actions the employer has taken or plans to take to control those hazards.

Exemption: Employers with fewer than 10 employees are exempt from the recordkeeping requirements of this rule.

(e) Hazard prevention and control.

(e)(1) What is the employer's basic obligation? The employer's basic obligation is to systematically comply with the hazard prevention and control requirements of the General Duty Clause and OSHA standards.

(e)(2) If it is not possible for the employer to comply immediately, what must the employer do? The employer must develop a plan for coming into compliance as promptly as possible, which includes setting priorities and deadlines and tracking progress in controlling hazards.

Note: Any hazard identified by the employer's hazard identification and assessment process that is covered by an OSHA standard or the General Duty Clause must be controlled as required by that standard or that clause, as appropriate.

(f) Information and training.

(f)(1) What is the employer's basic obligation? The employer must ensure that:

(i) Each employee is provided with information and training in the safety and health program; and
(ii) Each employee exposed to a hazard is provided with information and training in that hazard.

Note: Some OSHA standards impose additional, more specific requirements for information and training. This rule does not displace those requirements.

(f)(2) What information and training must the employer provide to exposed employees? The employer must provide information and training in the following subjects:

(i) The nature of the hazards to which the employee is exposed and how to recognize them;
(ii) What is being done to control these hazards;
(iii) What protective measures the employee must follow to prevent or minimize exposure to these hazards; and
(iv) The provisions of applicable standards.

(f)(3) When must the employer provide the information and training required by this rule?

(f)(3)(i) The employer must provide initial information and training as follows:

(A) For current employees, before the compliance date specified in paragraph (i) for this paragraph (f); and
(B) For new employees, before initial assignment to a job involving exposure to a hazard.

Note: The employer is not required to provide initial information and training in any subject in paragraph (f)(2) for which the employer can demonstrate that the employee has already been adequately trained.

(f)(3)(ii) The employer must provide periodic information and training:

(A) As often as necessary to ensure that employees are adequately informed and trained; and
(B) When safety and health information or a change in workplace conditions indicates that a new or increased hazard exists.

(f)(4) What training must the employer provide to employees who have program responsibilities? The employer must provide all employees who have program responsibilities with the information and training necessary for them to carry out their safety and health responsibilities.

(g) Evaluation of program effectiveness.

(g)(1) What is the employer's basic obligation? The employer's basic obligation is to evaluate the safety and health program to ensure that it is effective and appropriate to workplace conditions.

(g)(2) How often must the employer evaluate the effectiveness of the program? The employer must evaluate the effectiveness of the program:

(i) As often as necessary to ensure program effectiveness;
(ii) At least once within the 12 months following the final compliance date specified in paragraph (i); and
(iv) Thereafter at least once every two years.

(g)(3) When is the employer required to revise the program? The employer must revise the program in a timely manner to correct deficiencies identified by the program evaluation.

(h) Multi-employer workplaces.

(h)(1) What are the host employer's responsibilities? The host employer's responsibilities are to:

(i) Provide information about hazards, controls, safety and health rules, and emergency procedures to all employers at the workplace; and
(ii) Ensure that safety and health responsibilities are assigned as appropriate to other employers at the workplace.

(h)(2) What are the responsibilities of the contract employer? The responsibilities of a contract employer are to:

(i) Ensure that the host employer is aware of the hazards associated with the contract employer's work

> and what the contract employer is doing to address them; and
>
> (ii) Advise the host employer of any previously unidentified hazards that the contract employer identifies at the workplace.

practices in the operation of safety and health systems and a low incidence of occupational injuries and illnesses (OSHA 1998b). Where effective safety and health management is practiced, injury and illness rates are significantly less than rates at comparable work sites where safety and health management is weak or nonexistent. OSHA has developed several guidance documents, tools, and training to assist employers and safety professionals in the development and implementation of effective safety management systems. Sidebar 2 discusses a brief history of the OSHA consultative program.

In 1989, OSHA issued guidance that consists of safety and health management practices that are used by employers who are successful in protecting the safety and health of employees. The four major elements of an employer safety and health program identified by OSHA include (1) management, leadership, and employee involvement, (2) work-site analysis, (3) hazard prevention, and (4) safety and health training (OSHA). In January 2001 OSHA developed an eTool, a stand-alone, interactive, Web-based training tool on occupational safety and health payoffs (OSHA 2007b). The tool is highly illustrated and uses interactive graphical menus. OSHA also provides expert advisor software to employers and safety professionals to evaluate the financial impact of sound safety management programs. OSHA's $AFETY PAYS program is interactive software developed by OSHA to assist employers in assessing the impact of occupational injuries and illnesses (with lost work days) on their profitability (OSHA 1998c). It uses a company's profit margin, the average costs of an injury or illness, and an indirect cost multiplier to project the amount of sales a company would need to generate in order to cover those costs.

In 1982, OSHA began recognizing those employers that voluntarily participated in promoting effective

SIDEBAR 2

(Source: OSHA 2001a. Directive 00-01 (CSP 02), TED 3.6, *Consultation Policies and Procedures Manual, Chapter 1, Section IX: A Brief History of the OSHA Consultation Program.* (August 6, 2001).

A Brief History of the OSHA Consultation Program.

Section 21(c) of the Occupational Safety and Health Act of 1970 (the Act) requires the Secretary of Labor to establish programs for the education and training of employers and employees in recognizing, avoiding, and preventing unsafe or unhealthful working conditions covered under the Act. Many States began providing onsite consultation services to employers as part of their State plan under Section 18(b) of the Act. OSHA soon recognized that employers needed help in understanding and complying with the sometimes complex regulations applying to their workplaces. In addition, small employers often lack the financial resources to hire private consultants to aid them in meeting their obligations under the Act. In response to the demand for similar onsite consultation in Federal enforcement States, in 1975, the Secretary of Labor set forth the regulation at 29 CFR Part 1908 (FR 40: 21935), which authorized Federal funding of onsite consultation activity by States under Federal OSHA's jurisdiction. This activity was funded through Cooperative Agreements under the authority of Sections 21(c) and 7(c)(1) of the Act.

In 1977, the level of Federal funding for State run consultation projects was increased to ninety percent, a level that provided a strong incentive for all States to enter into the program. Forty-eight States, the District of Columbia, and Guam operate OSHA onsite consultative programs under Section 21(d) agreements with Federal OSHA. Two States and two U.S. territories operate programs as part of their approved State plans, for which fifty percent funding is received from Federal OSHA through 23(g) grants.

Part 1908 has been amended several times in the intervening years. In 1983, OSHA published a proposed change to the consultation regulation to clarify a number of provisions and to change the focus of services provided to an employer during an OSHA consultative visit. The proposal raised a number of new issues, including the Agency's desire to shift the focus of the consultation visit from simply the identification and correction of specific workplace hazards to the broader and more comprehensive goal of addressing the employer's overall management system for ensuring a safe and healthful workplace. In addition, the proposal allowed for offsite consultation services, including training and education services, to be made available to employers. It also provided for an exemption from Programmed OSHA Inspections for employers who met specific criteria. A final rule including these provisions was published in the Federal Register on June 19, 1984 (FR 49: 25082).

The Occupational Safety and Health Compliance Assistance Authorization Act of 1998, Public Law 105-197, codified OSHA's Consultation Program and amended Section 21 of the OSH Act by adding a new subsection, (d). On October 26, 2000, 29 CFR Part 1908 was amended to ensure that employees would be allowed to participate in site visits, that employees would be informed of the results of site visits, that site visits would be conducted according to updated procedures, and that information obtained during site visits would be treated as confidential.

I. *How the Consultation Program Works.*
   A. The consultation program is designed to assist employers in identifying and correcting serious hazards in the workplace. Priority in scheduling visits is generally given to small employers in high hazard industries. Consultation projects also provide assistance to employers in developing safety and health management systems. However, this assistance must be linked to a hazard evaluation visit by either the consultation project, by OSHA enforcement, or by a private consultant. The consultation project must have access to the report of the visit before providing program assistance. In the case of offsite technical training, the Consultation Project Manager may provide specific training services that are not directly related to an onsite visit.
   B. Because consultation services are voluntary, an employer must request service and agree to certain obligations, the principal one being that the employer agrees to correct all serious hazards found during the consultation visit within an agreed-upon timeframe.

work-site-based safety and health management programs through their Voluntary Protection Programs (VPP). In the VPP, management, labor, and OSHA establish cooperative relationships at workplaces that have implemented a comprehensive safety and health management system. Acceptance into VPP is OSHA's official recognition of the outstanding efforts of employers and employees who have achieved exemplary occupational safety and health (OSHA 2004). The legislative underpinning for VPP is found in Section (2)(b)(1) of the Occupational Safety and Health Act of 1970, which declares the Congress's intent "to assure so far as possible every working man and woman in the Nation safe and healthful working conditions and to preserve our human resources—(1) by encouraging employers and employees in their efforts to reduce the number of occupational safety and health hazards at their places of employment, and to stimulate employers and employees to institute new and to perfect existing programs for providing safe and healthful working conditions" (OSHA 2004).

In practice, VPP sets performance-based criteria for a managed safety and health system, invites sites to apply, and then assesses applicants against these criteria. OSHA's verification includes an application review and a rigorous on-site evaluation by a team of OSHA safety and health experts. OSHA approves admission of qualified sites to one of three programs: Star, Merit, and Star Demonstration. Star Demonstration recognizes work sites that address unique safety and health issues. As of February 2005, there were over 1250 participants in both state and federal VVP plans. OSHA has also developed a 4-day training course to increase the understanding of safety and health management principles for all OSHA enforcement personnel, State Plan consultants, and other interested stakeholders. This course emphasizes the VPP culture, philosophy, and criteria as the basis for students to understand and evaluate any safety and health management systems. Currently this course is offered at the OSHA Training Institute four times a year.

Elements of an Effective Safety Management System

Effective safety management involves developing and implementing several operational elements and integrating them into the organization's management

SIDEBAR 3

Safety and Health Program Management Guidelines; Issuance of Voluntary Guidelines OSHA Federal Register Notice 54:3904-3916, January 26, 1989

The Guidelines

(A) General. (1) Employers are advised and encouraged to institute and maintain in their establishments a program which provides systematic policies, procedures, and practices that are adequate to recognize and protect their employees from occupational safety and health hazards.

(2) An effective program includes provisions for the systematic identification, evaluation, and prevention or control of general workplace hazards, specific job hazards, and potential hazards which may arise from foreseeable conditions.

(3) Although compliance with the law, including specific OSHA standards, is an important objective, an effective program looks beyond specific requirements of law to address all hazards. It will seek to prevent injuries and illnesses, whether or not compliance is at issue.

(4) The extent to which the program is described in writing is less important then how effective it is in practice. As the size of a worksite or the complexity of a hazardous operation increases, however, the need for written guidance increases to ensure clear communications of policies and priorities and consistent and fair application of rules.

(B) Major Elements. An effective occupational safety and health program will include the following four elements. To implement these elements, it will include the actions described in paragraph (C).

(1) Management commitment and employee involvement are complementary. Management commitment provides the motivating force and the resources for organizing and controlling activities within an organization. In an effective program, management regards workers safety and health as a fundamental value of the organization and applies its commitment to safety and health protection with as much vigor as to other organizational purposes. Employee involvement provides the means through which workers develop and/or express their own commitment to safety and health protection, for themselves and for their fellow workers.

(2) Worksite analysis involves a variety of worksite examinations, to identify not only existing hazards but also conditions and operations in which changes might occur to create hazards. Unawareness of a hazard which stems from failure to examine the worksite is a sure sign that safety and health policies and/or practices are ineffective. Effective management actively analyzes the work and worksite, to anticipate and prevent harmful occurrences.

(3) Hazard prevention and controls are triggered by a determination that a hazard or potential hazard exists. Where feasible, hazards are prevented by effective design of the jobsite or job. Where it is not feasible to eliminate them, they are controlled to prevent unsafe and unhealthful exposure. Elimination or controls is accomplished in a timely manner, once a hazard or potential hazard is recognized.

(4) Safety and health training addresses the safety and health responsibilities of all personnel concerned with the site, whether salaried or hourly. If is often most effective when incorporated into other training about performance requirements and job practices. Its complexity depends on the size and complexity of the worksite, and the nature of the hazards and potential hazards at the site.

(C) Recommended Actions

(1) Management Commitment and Employee Involvement.

(i) State clearly a worksite policy on safe and healthful work and working conditions, so that all personnel with responsibility at the site and personnel at other locations with responsibility for the site understand the priority of safety and health protection in relation to other organizational values.

(ii) Establish and communicate a clear goal for the safety and health program and objectives for meeting that goal, so that all members of the organization understand the results desired and the measures planned for achieving them.

(iii) Provide visible top management involvement in implementing the program, so that all will understand that management's commitment is serious.

(iv) Provide for encouragement of employee involvement in the structure and operation of the program and in decisions that affect their safety and health, so that they will commit their insight and energy to achieving the safety and health program's goal and objectives.

(v) Assign and communicate responsibility for all aspects of the program so that managers, supervisors, and employees in all parts of the organization know what performance is expected of them.

(vi) Provide adequate authority and resources to responsible parties, so that assigned responsibilities can be met.
(vii) Hold managers, supervisors, and employees accountable for meeting their responsibilities, so that essential tasks will be performed.
(viii) Review program operations at least annually to evaluate their success in meeting the goal and objectives, so that deficiencies can be identified and the program and/or the objectives can be revised when they do not meet the goal of effective safety and health protection.

(2) Worksite Analysis. (i) So that all hazards are identified:
(a) Conduct comprehensive baseline worksite surveys for safety and health and periodic comprehensive update surveys;
(b) Analyze planned and new facilities, processes, materials, and equipment; and
(c) Perform routine job hazard analyses.
(ii) Provide for regular site safety and health inspection, so that new or previously missed hazards and failures in hazard controls are identified.
(iii) So that employee insight and experience in safety and health protection may be utilized and employee concerns may be addressed, provide a reliable system for employees, without fear of reprisal, to notify management personnel about conditions that appear hazardous and to receive timely and appropriate responses; and encourage employees to use the system.
(iv) Provide for investigation of accidents and "near miss" incidents, so that their causes and means for their prevention are identified.
(v) Analyze injury and illness trends over time, so that patterns with common causes can be identified and prevented.

(3) Hazard Prevention and Control. (i) So that all current and potential hazards, however detected, are corrected or controlled in a timely manner, establish procedures for that purpose, using the following measures:
(a) Engineering techniques where feasible and appropriate;
(b) Procedures for safe work which are understood and followed by all affected parties, as a result of training, positive reinforcement, correction of unsafe performance, and, if necessary, enforcement through a clearly communicated disciplinary system;
(c) Provision of personal protective equipment; and
(d) Administrative controls, such as reducing the duration of exposure.

(ii) Provide for facility and equipment maintenance, so that hazardous breakdown is prevented.
(iii) Plan and prepare for emergencies, and conduct training and drills as needed, so that the response of all parties to emergencies will be "second nature."
(iv) Establish a medical program which includes availability of first aid on site and of physician and emergency medical care nearby, so that harm will be minimized if any injury or illness does occur.

(4) Safety and Health Training. (i) Ensure that all employees understand the hazards to which they may be exposed and how to prevent harm to themselves and others from exposure to these hazards, so that employees accept and follow established safety and health protections.
(ii) So that supervisors will carry out their safety and health responsibilities effectively, ensure that they understand those responsibilities and the reasons for them, including:
(a) Analyzing the work under their supervision to identify unrecognized potential hazards;
(b) Maintaining physical protections in their work areas; and
(c) Reinforcing employee training on the nature of potential hazards in their work and on needed protective measures, through continual performance feedback and, if necessary, through enforcement of safe work practices.
(iii) Ensure that managers understand their safety and health responsibilities, as described under (C)(1). "Management Commitment and Employee Involvement," so that the managers will effectively carry out those responsibilities.

structure. The safety management system should go beyond mere regulatory compliance. Sidebar 3 presents OSHA's proposed safety and health program rule from 1989. This proposed compliance document was geared toward addressing OSHA's general duty clause.

Sidebar 4 presents an outline of the ANSI Standard Z10-2005, *Occupational Health and Safety Management Systems*. A review of each program will illustrate the major differences between a safety compliance program and a safety management system. This section addresses how to develop and implement each of these elements of a successful and effective safety management system.

SIDEBAR 4

ANSI Z10, Occupational Health and Safety Management Systems Table of Contents

(ANSI/AIHA 2005)

1.0 Scope, Purpose, and Application
1.1 Scope
1.2 Purpose
1.3 Application
2.0 Definitions
3.0 Management Leadership and Employee Participation
3.1 Management Leadership
3.1.1 Occupational Health and Safety Management system
3.1.2 Policy
3.1.3 Responsibility and Authority
3.2 Employee Participation
4.0 Planning
4.1 Initial and Ongoing Review
4.1.1 Initial Review
4.1.2 Ongoing Review
4.2 Assessment and Prioritization
4.3 Objectives
4.4 Implementation Plans and Allocation of Resources
5.0 Implementation and Operation
5.1 OHSMS Operation Elements
5.1.1 Hierarchy of Controls
5.1.2 Design Review and Management of Change
5.1.3 Procurement
5.1.4 Contractors
5.1.5 Emergency Preparedness
5.2 Education, Training, and Awareness
5.3 Communication
5.4 Documentation and Record Control Process
6.0 Evaluation and Corrective Action
6.1 Monitoring and Measurement
6.2 Incident Investigation
6.3 Audits
6.4 Corrective and Preventive Actions
6.5 Feedback to the Planning Process
7.0 Management Review
7.1 Management Review Process
7.2 Management Review Outcomes and Follow Up

Annexes

A Policy Statements (Section 3.1.2)
B Roles and Responsibilities (Section 3.1.3)
C Employee Participation (Section 3.2)
D Initial/Ongoing Review (Section 4.1)
E Assessment and Prioritization (Section 4.2)
F Objectives/Implementation Plans (Section 4.3 and 4.4)
G Hierarchy of Control (Section 5.1.1)
H Incident Investigation Guidelines (Section 6.2)
I Audit (Section 6.3)
J Management Review Process (Section 7.1 and 7.2)
K Bibliography and References

The 11 annexes provide explanatory comments, examples of forms and procedures, and reference sources for many of the major sections. While information in the annexes is not part of the standard, it will be helpful to those assigned responsibility to implement the standard.

Management Leadership and Employee Involvement

Management leadership and employee involvement go hand in hand for safety success. In fact, top management leadership and effective employee participation are crucial for the success of a safety management system (ANSI/AIHA 2005). Management provides the leadership for organizing and controlling activities within an organization. They provide the motivating force, resources, and influence necessary to place safety as a fundamental value within the organization. In an effective program, management involvement also provides the means through which workers express their own commitment to safety and health for themselves and their fellow workers (OSHA 1989).

The ANSI Z10 standard identifies management leadership as the first step of a successful safety management system (ANSI/AIHA 2005). The voluntary consensus standard identifies management's involvement as follows:

- Top management shall direct the organization to establish, implement, and maintain an Occupational Health and Safety Management System (OHSMS).
- The organization's top management shall establish a documented occupational health and safety policy.

- Top management shall provide leadership and assume overall responsibility.
- The organization shall establish and implement processes to ensure effective participation in the OHSMS by its employees at all levels.

The ANSI Z10 standard also addresses employee involvement in an effective safety management system by stating, "Employees shall assume responsibility for aspects of health and safety over which they have control, including adherence to the organization's health and safety rules and requirements (ANSI/AIHA 2005).

Petersen (2003) developed a list of principles of safety management that illustrates the roles of the employee and management in effective safety management. First, the safety system should fit into the culture of the organization. Next, safety should be managed like any other company function, with strong leadership. In most cases, unsafe behavior is normal human behavior, and an unsafe act, unforeseen condition, or an accident are all symptoms that something is wrong with the organization's management system. It is therefore management's job to change the environment that leads to the unsafe behavior or hazard present (Petersen 2003). The safety professional must facilitate this process by identifying the root causes of accidents or occupational hazards so that the solutions work toward changing the organization's management culture. This requires a flexible process involving the visibility of top management, involvement of middle management, accountability of supervisors, and employee involvement. The SHE professional should also implement these programs so that they are perceived as a positive impact to the culture of the organization. This ambitious undertaking begins very simply with an assessment of the organization's culture.

Management, Culture, and Safety

The leadership of an organization influences its overall culture. The organization's culture in turn "determines what will—and will not—work in SHE efforts" (Petersen 2004, 28). The culture of an organization influences the attention paid to job-site hazards, risks, and the safety attitudes of the employees. This safety culture must be evaluated so the SHE professional can develop a strategy for: (1) improving management commitment to safety, (2) establishing realistic goals of where the organization needs to be, (3) preparing achievable objectives to effectively meet those goals, and (4) developing metrics to measure the safety management system's success toward achieving the stated goals.

Several surveys of the SHE community have consistently cited integrating safety into the company culture as a top priority for safety professionals (Hintch 2005, ASSE and BCSP 2007a). Integrating safety into the culture of an organization requires leadership from management. However, management must be committed to safety before they can lead. Obtaining the commitment of upper management to safety management requires the SHE professional to evaluate the existing culture of the organization and its attitude toward safety. SHE professionals may find that management's attitude toward safety falls into one of the following categories:

- Management is already committed to safety. Safety is an investment with a measurable return.
- Management is aware of some of the benefits of safety but does not lead. Safety pays for itself or is an acceptable cost.
- Management is aware of safety compliance but does not understand safety benefits. Safety is compliance-driven. Safety is a cost to the organization.
- Management is not committed to safety. Production and profits take precedence over workplace safety and health. Safety negatively impacts production. Hazards and accidents are part of doing business.

These are but a few of the attitudes held by management that affect the safety culture in an organization. There are many other variations in an organization's culture and how it impacts on safety.

The key for the SHE professional is to use appropriate tools to determine where the current culture of the organization is, compared to where it needs to be.

The SHE professional should not underestimate management's understanding and commitment to safety. A recent survey of 500 CEOs on awareness of safety benefits by the American Society of Safety Engineers (ASSE 2003) found:

- 60 percent said safety contributes to profits
- 22 percent said safety pays for itself
- 10 percent called safety a minor expense
- 5 percent called safety a major expense
- 3 percent did not respond

Although this study was limited, it provides anecdotal findings that four out of five CEOs see safety as breaking even or contributing to profits. Another recent publication provided insight into the successful companies that "get" safety (Colford 2005b). One key issue found in successful companies was that they embrace safety as a core value. These companies also found safety is nonnegotiable, uncompromising, and permanent. Finally, the cultures of these successful companies indicated that providing a safe workplace is the right thing to do regardless of financial payback.

The SHE professional will have to use a variety of tools to evaluate where their company's culture is to achieve the goals of successful companies such as those cited above.

The SHE professional must use other tools to continue to demonstrate to management that safety provides a return on investment for the company and should be incorporated into the business plan like all other departments in the organization. Tools of the trade for a successful safety management system will be discussed later in this chapter.

Employee Involvement

Employee involvement in the safety management system is essential for its success. We often hear the phrase, "Safety is everyone's responsibility." Of course, safety is everyone's responsibility, but what does this responsibility mean?

First, everyone is responsible for safety to some degree, including employees. Employees who are accountable for everyday safety must be able to contribute to reasonable and achievable safety program elements. Teamwork has a lot to do with a company's culture. It is hard to have teamwork when you are not part of the team. Therefore, it is essential for employees to be involved in the safety management system and to effect a positive change in workplace cultre.

Secondly, employees provide critical insight into workplace- and job task-related hazards. Employees see the implementation and effectiveness of safety objectives firsthand. This awareness and feedback is also important to the safety program's success. Special emphasis should be given to participation by nonsupervisory employees because they are often those closest to the hazard, and often have the most intimate knowledge of workplace hazards. In addition, nonsupervisory employees are a valuable but often overlooked resource for improving health and safety (ANSI/AIHA 2005).

Third, management and the SHE professional cannot observe all operations all of the time. Developing a strong culture with employee involvement provides peer observation and self-observation to follow established safety principles "when no one is looking."

Fourth, employees have legal and regulatory obligations to act in a safe manner. Employers' actions can affect legal and regulatory obligations of the company as well.

Finally, employees are generally involved with contributing to accidents and injuries and/or suffering the consequences of them. Employees are involved in the safety management system and the organization's culture, whether it is acknowledged by management or not. The key is whether employee involvement contributes positively or negatively to the overall culture of an organization. The SHE professional must involve employees so that there is a positive effect on the overall safety culture of the organization.

There are multiple ways to get employees involved in the safety management system. Many authors (Anton 1989, Della-Giustina 2000, Petersen 2003, Reese 2003) suggest examples of employee participation, including:

- participating on safety committees
- conducting work-site safety inspections
- performing a job-site hazard analysis identifying routine hazards in each step of a job or process
- preparing safe work practices or controls to eliminate or reduce exposure to job hazards
- suggesting revisions and improvements to safety and health rules
- participating during safety training
- providing programs and presentations at safety and health meetings
- participating in or assisting with incident investigations
- reporting recognized workplace hazards and unsafe work practices
- recommending corrective actions to fix hazards within one's control
- supporting fellow workers by providing feedback on risks and assisting them in eliminating hazards
- suggesting or performing a pre-use or change analysis for new equipment or processes in order to identify hazards before use

Employees' active involvement is mandatory if safety is to be integrated into an organization's culture. Leadership from management is the other component for a successful management system. The success of the safety management system is a sum of how both of these two groups communicate as a whole. The SHE professional plays a critical role in facilitating management leadership and employee involvement and buy-in to the values and benefits of workplace safety (see Sidebar 5).

Work-Site Analysis

As previously discussed, a primary goal for the safety management system is to prevent and minimize

---

SIDEBAR 5

Evaluating Management Leadership and Employee Involvement (OSHA 1996)

According to OSHA, an organization will have successfully established management leadership and employee involvement when it can demonstrate:

Work Site Safety and Health Program

- There is a written policy supported by senior management that promotes safety and health.
- The policy is straightforward and absolutely clear.
- The policy can be easily explained or paraphrased by the entire workforce.
- The policy is expressed in the context of other organizational values. It goes beyond compliance to address the safety behavior of all members of the organization and guides them in making
- Decisions in favor of safety and health when apparent conflicts arise with other values and priorities.

Setting and Communicating Clear Goals and Objectives

- A safety and health goal directly related to the safety and health policy exists in writing.
- The goal incorporates the essence of a positive and supportive safety system integrated into the workplace culture.
- An assessment tool is used to identify deficiencies that relate to objectives that are designed to achieve the goal, and they are assigned to responsible individuals.
- A measurement system is consistently used to manage work on objectives and indicate progress towards the goal.
- The goal and objectives are supported by senior management and can be paraphrased by the entire workforce.
- Measures used to track objective progress are known to the workforce and members of the workforce are active participants in the objective process.

Management Leadership

- The positive influence of management is evident in all elements of the safety and health program.

- All members of the workforce perceive management to be exercising positive leadership.
- All members of the workforce can give examples of management's positive leadership.

Management Sets Examples

- All managers know and understand the safety and health rules of the organization and the safe behaviors they expect from others.
- Managers throughout the organization consistently follow the rules and behavioral expectations set for others in the workforce as a matter of personal practice.
- Members of the workforce perceive management to be consistently setting positive examples and can illustrate why they hold these positive perceptions.
- A majority of members of management at all levels consistently address the safety behavior of others by coaching and correcting poor behavior and positively reinforcing good behavior.
- Members of the workforce credit management with establishing and maintaining positive safety values in the organization through their personal example and attention to the behavior of others.

Employee Involvement

- Employees accept personal responsibility for ensuring a safe and healthy workplace.
- The employer provides opportunities and mechanisms for employees to influence safety and health program design and operation.
- There is evidence of management support of employee safety and health interventions.
- All employees have a substantial impact on the design and operation of the safety and health program.
- There are multiple avenues for employee participation.
- The avenues are well known, understood, and used by all employees.
- The avenues and mechanisms for involvement are effective at reducing accidents and enhancing safe behaviors.

jobsite hazards. Therefore, the hazards must be identified and subsequent risks assessed in order to select appropriate control options to address them (Manuele 2005). A work-site analysis means that managers and employees analyze all work-site conditions to identify and eliminate any existing or potential hazards.

There should be a comprehensive baseline survey, with a system in place for periodic updates. However, the SHE professional should ensure that the staff and resources are available to address any identified hazards promptly before initiating the program.

To help in conducting a work-site analysis, OSHA suggests the following:

- Request a free OSHA consultation visit.
- Become aware of hazards in your industry.
- Create safety teams.
- Encourage employees to report hazards.
- Provide an adequate system for reporting hazards.
- Encourage trained personnel to conduct inspections of the work site and correct hazards.
- Ensure that any changes in process or new hazards are reviewed.
- Seek assistance from safety and health experts.

There are many ways to conduct a work-site analysis. OSHA suggests the following plan to identify work-site hazards (see Figure 2):

- Conduct a comprehensive, baseline survey for safety and health and periodic, comprehensive update surveys.
- Change analysis of planned and new facilities, processes, materials, and equipment.
- Perform routine job-hazard analyses.
- Conduct periodic and daily safety and health inspections of the workplace.

These analytical tools are used to identify not only existing hazards, but also conditions and operations in which changes might create new hazards. An organization's management team can demonstrate its commitment to safety through leadership that actively analyzes job tasks and the work site to anticipate, prevent, or address workplace hazards before they occur.

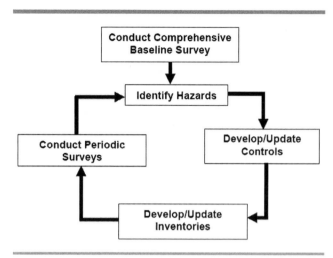

FIGURE 2. OSHA work-site analysis chart (*Source:* OSHA 2001a)

There are many hazards in a workplace that the SHE professional should make sure are evaluated during any analysis of the workplace. Some of the more common workplace hazards include chemicals (toxic, flammable, corrosive), explosions (chemical reactions, over-pressurizations), electricity (shock, short circuit, static discharge, loss of power, fire), ergonomics (human errors, strains), excavation (collapse/cave-in), falls (slips, trips, from heights), fire, heat, vibration (chaffing, fatigue), machinery (failure, crushing, caught between), noise, radiation (ionizing, nonionizing), struck by or against, temperature extremes, poor visibility, and weather (OSHA 2005). The ANSI Z10 standard discusses hazards and risk by stating, "Hazards and risks may include health, medical and weather-related emergencies, as well as emergency events that may arise from the characteristics of the materials, processes and activities of the workplace" (ANSI/AIHA 2005, 15).

There are also certain situations that can contribute to the severity of accidents and injuries. Petersen (2004) found that a number of recent studies suggested that severe injuries are fairly predictable in certain situations, including:

- unusual nonroutine work, including unique, one-of-a-kind projects where normal routine controls have little effect
- nonproduction activities, including maintenance functions and research and development activities that may not be carried out by standardized procedures or receive little safety attention
- sources of high energy, including electricity, steam, compressed gases, and flammable liquids
- unique construction activities, including steel erection, tunneling, and work over water where unusual risks are present
- various lifting situations that result in the large and costly problem of severe strains and injuries, primarily to workers' backs
- repetitive motion activities that lead to chronic conditions and costly surgeries involving tendonitis and carpal tunnel syndrome
- psychological stress situations resulting from employee exposures to stressful environments
- exposure to toxic materials that can have acute or chronic health problems that may become problems for decades, as in the example of asbestos

Hazard Prevention

It is important to take steps to eliminate or control the uncontrolled hazards that were identified during workplace surveillance and reduce them to an acceptable risk level. Therefore, these hazard prevention methods are only going to be as effective as the hazard-assessment *and* risk-evaluation process. Once hazards are identified, the SHE professional must perform some type of risk evaluation of each hazard to define its relationship to worker exposure (including frequency and duration), severity of consequences, and probability of occurrences. Then the SHE professional can rank the risks, develop a hazard remediation strategy, and take action.

The ANSI Z10, *Standard for Occupational Health and Safety Management Systems*, discusses hazard assessment and prioritization by stating, "The assessment of risks should include factors such as identification of potential hazards, exposure, measurement data, sources and frequency of exposure, types of measures used to control hazards, and potential severity of hazards. Assessing risks can be done using quantitative (numeric) or qualitative (descriptive) methods.

There are many methods of risk assessment" (ANSI/AIHA 2005). Some of these risk-assessment methods include:

- Preliminary Hazard Analysis (PHA)
- Safety Reviews—Operations Analyses
- What-If Analysis
- Checklist Analysis
- What-If/Checklist Analysis
- Hazard and Operability Analysis (HAZOP)
- Failure Modes and Effects Analysis (FMEA)
- Fault Tree Analysis (FTA)

Risk-assessment methodologies are discussed in later chapters of this manual.

The SHE professional should also be aware that since any actual knowledge of hazards that constitute OSHA violations can be discovered in the course of a compliance investigation or the litigation process, documenting proper control of identified hazards is important. This actual or imputed knowledge can be the basis for willful violations, which can carry significant penalties and—if a fatality was involved—form the basis for a criminal referral by OSHA to the U.S. Department of Justice under Section 17 of the OSH Act. If a nonemployee is harmed, the OSHA citations will come in as "negligence per se" (considered intrinsically negligent) to substantiate the plaintiff's tort injury or wrongful death action. In some states, a high negligence (willful) OSHA citation permits an employee to get around the workers' compensation shield (Abrams 2005). A proactive safety hazard-analysis program will address identified hazards immediately. Figure 3 displays the four major actions that form the basis from which good hazard prevention and control can develop.

The SHE professional should ensure that the staff and resources are available to address any identified hazards promptly before initiating the hazard recognition program. It is impractical to attempt to achieve an immediately hazard-free environment due to the excessive costs and resources that would be necessary. However, the total abatement of all identified hazards does result in total compliance, regardless of the costs. There will have to be some decisions made on what hazards to address, then prioritize their importance. What resources should be allocated and what is the return on investment or cost savings to be achieved? Financial and management decisions that address hazards will be prioritized along with all other organizational business decisions.

The following is a summary of a hazard-analysis and risk-assessment guide developed by Fred A. Manuele (2005):

1. Establish analysis parameters. Select a manageable task, system, process, or product to be analyzed, and establish its boundaries and operating phase. Determine the scope of analysis in terms of what can be harmed or damaged: people, property, equipment, productivity, and the environment.
2. Identify the hazards. The frame of thinking adopted should get to the root of causal factors, which are hazards. These questions need to be asked: What are the characteristics of things or the actions or inactions of people that present a potential for harm?
3. Consider the failure modes. Define possible failure modes that would result in the realization of the potential of the hazards. Consider how an undesirable event could occur and what controls are in place to mitigate its occurrence.
4. Determine exposure frequency and duration. For each harm or damage category selected in Step 1 for the scope of the analysis, estimate the frequency and duration of exposure to the hazard.
5. Assess the severity of consequences. What is the magnitude of harm or damage that could result? Learned speculations must be made regarding the consequences of occurrence: the number of resulting injuries and illnesses or fatalities, the value of property or equipment damaged, the duration of lost productivity, or the extent of environmental damage. Historical data can establish a baseline. When severity of consequences is determined, the hazard analysis is complete.

These four major actions form the basis from which good hazard prevention and control can develop.

Comprehensive surveys

For small businesses, OSHA-funded, state-run consultation services can conduct a comprehensive survey at no cost. Many workers' compensation carriers and other insurance companies offer expert services to help their clients evaluate safety and health hazards. Numerous private consultants provide a variety of safety and health expert services. Larger businesses may find the needed expertise at the company or corporate level.

For the industrial hygiene survey, at a minimum, all chemicals and hazardous materials in the plant should be inventoried, the hazard communication program should be reviewed, and air samples analyzed. For many industries, a survey of noise levels, a review of the respirator program, and a review of ergonomic risk factors are needed.

Change Analysis

Anytime something new is brought into the workplace, whether it be a piece of equipment, different materials, a new process, or an entirely new building, new hazards may unintentionally be introduced. Before considering a change for a worksite, it should be analyzed thoroughly beforehand. Change analysis helps in heading off a problem before it develops.

You may find change analysis useful when:

- Building or leasing a new facility.
- Installing new equipment.
- Using new materials.
- Starting up new processes.
- Staffing changes occur.

Hazard Analysis

Hazard analysis techniques can be quite complex. While this is necessary in some cases, frequently a basic, step-by-step review of the operation is sufficient. One of the most commonly used techniques is the Job Hazard Analysis (JHA). Jobs that were initially designed with safety in mind may now include hazards or improper operations. When done for every job, this analysis periodically puts processes back on the safety track.

Other, more sophisticated techniques are called for when there are complex risks involved. These techniques include: WHAT-IF Checklist, Hazard and Operability Study, Failure Mode and Effect Analysis, and Fault Tree Analysis.

Safety and Health Inspections

Routine site safety and health inspections are designed to catch hazards missed at other stages. This type of inspection should be done at regular intervals, generally on a weekly basis. In addition, procedures should be established that provide a daily inspection of the work area.

You can use a checklist already developed or make your own, based on:

- Past problems.
- Standards that apply to your industry.
- Input from everyone involved.
- Your company's safety practices or rules.

Important things to remember about inspections are:

- Inspections should cover every part of the worksite.
- They should be done at regular intervals.
- In-house inspectors should be trained to recognize and control hazards.
- Identified hazards should be tracked to correction.

Information from inspections should be used to improve the hazard prevention and control program.

FIGURE 3. Four major actions from which good hazard prevention and control develop (*Source:* OSHA 2001a)

6. Determine occurrence probability. Consider the likelihood that a hazardous event will occur. This process is also subjective. Probability must be related to an interval base of some sort, such as a unit of time, an activity, events, units produced, or the life cycle of a facility, equipment, process, or product.
7. Define the risk. Conclude with a statement that addresses both the probability of an incident occurring and the expected severity of harm or damage. Categorize each risk in accord with agreed-upon terms, such as high, serious, moderate, or low.
8. Rank risks in priority order. Risks should be ranked in order to establish priorities. Since the hazard-analysis and risk-assessment exercise is subjective, the risk-ranking system will also be subjective.
9. Develop remediation proposals. When required by the results of the risk assessment, alternate proposals for design and operational changes that are needed to achieve an acceptable risk level would be recommended.
10. Actions and follow-up activities. Actions should be taken as necessary, as should follow-up activities to determine whether the action was effective.

Although each organization will address hazard abatement differently, certain steps should be considered to handle them in a prioritized manner. Petersen outlines four general principles for conducting hazard-abatement activities (Petersen 2003).

1. Set priorities. Most organizations cannot do everything at once, so decisions must be made about what comes first. Priorities can be determined based upon cost, severity of the hazard, risk, regulatory violations, or other ranking systems.
2. Schedule and assign tasks. Without planning and assigning responsibilities, nothing very significant will happen. Establishing deadlines for completing assigned corrective actions will create accountability for addressing identified hazards.
3. Follow up to ensure that things are being done. Hazard-abatement activities should be evaluated to ensure that the hazard reduction assignment has been performed correctly and in a timely manner. The hazard can be reevaluated to determine if the fix actually resulted in a reduction or elimination of the hazard. Also, a reevaluation can determine if a new hazard was created by the fix.
4. Document everything that is done, if not for internal reasons, certainly for OSHA. Documentation provides a written record that hazards are identified and that corrective actions are taken to abate the hazard.

Prioritizing hazard-abatement activities will vary based upon the types of hazards present, abatement costs (direct and indirect), complexity of the abatement method selected, and risk tolerance of the organization. For instance, replacing a burnt-out light bulb in an exit sign is a fairly simple and inexpensive hazard-abatement choice that can be implemented quite easily. The bulb replacement can be scheduled through daily work orders. A quick follow-up inspection will verify the exit light is properly illuminated. Finally, documenting the corrective action demonstrates the effectiveness of the employer's safety management system to find and abate workplace hazards. More expensive hazard abatement, such as designing and installing a guard system on machinery, might be a high priority if an insurance premium increases as long as the hazard exists. In this case, prompt scheduling, along with documented hazard abatement provided to the insurance carrier, may result in reduced premiums. However, if after prioritization an identified hazard cannot be fixed immediately, it should be isolated through securing the hazard area, tagging out the hazardous equipment, training employees on hazard avoidance, or using other remedial actions that are available. Most importantly, the SHE professional should document all interim actions in order to show that all reasonable steps were taken to protect workers until a permanent fix is completed.

It is critical that the SHE professional also ensure that the hazard prevention methods selected actually eliminate or reduce hazards and associated risks to acceptable levels. Judging the amount of risk deemed as acceptable will have an impact on defining the extent of injury and incident reduction. Judgment can be quantified if criteria are developed with values placed upon them.

Reese (2003, 106) suggests that one way to develop quantifiable judgments of risk is to calculate them using the following formula:

$$\text{Risk-Assessment Factor} = \text{Consequence} \times \text{Exposure} \times \text{Probability} \quad (1)$$

A point system is assigned to each of the sub-choices in each of the three areas used to calculate the risk-assessment factor.

- Consequences could be listed as multiple deaths (10 points), single death (9 points), multiple injuries (7 points), disabling injuries (6 points), serious injuries (5 points), other injuries (3 points), or first aid (1 point).
- Exposure could be defined as continuous (10 points), hourly (8 points), daily (6 points), weekly (4 points), and so on.
- Probability is expressed as a percentage, such as 100 percent (10 points), 75 percent (8 points), 50 percent (6 points), 25 percent (4 points), and 0 percent (2 points).

A risk-assessment factor can be developed based upon multiplying the values selected in consequences, exposures, and probability. A range of 801–1000 points might be deemed the highest risk. A range of 601–800 might be a high risk, and so on. This method can be customized for specific hazards or specific situations. The results can be used, not used, or combined with other risk-assessment tools. Cost justifications can then be performed to determine if the cost is worth the amount of hazard that is abated. This includes a determination of whether the cost of the hazard prevention activity had complete, partial, or no effect on resolving the intended hazards.

Good judgment for assessing risks of hazards involves understanding that the total human and economic burdens of occupational injuries and illnesses is crucial to setting priorities and shaping other components of the safety management plan. Furthermore, determining the magnitude of these economic burdens is essential to the assessment of the financial effectiveness of safety and health interventions designed to reduce the number of occupational injuries and illnesses. Such evaluations provide decision makers in organizations with necessary information to assess whether the outcomes of interventions justify the expenditures relative to other choices. Economic analysis is vital in preventing and controlling occupational injury and illness (NIOSH 2005). There should also be an evaluation of whether the hazard-prevention actions actually created new hazards during the process (Manuele 2005).

The selection of hazard abatement must follow an order of precedence to be effective in addressing and controlling the hazard. Figure 4 depicts the safety decision hierarchy.

The hierarchy of addressing hazards is summarized as follows:

1. Engineering Controls. The most effective control is by engineering out the hazard through physically changing a machine or work environment to prevent employee exposure to the hazard. Examples include:

A) **Problem Identification & Analysis**
  1) Identify and analyze hazards.
  2) Assess risks.

B) **Consider These Actions in Order of Effectiveness**

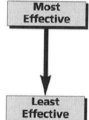

  1) Eliminate hazards and risks through system design and redesign.
  2) Reduce risks by substituting less-hazardous methods or materials.
  3) Incorporate safety devices.
  4) Provide warning systems.
  5) Apply administrative controls (e.g., work methods, training).
  6) Provide PPE.

C) **Decide & Take Action**

D) **Measure for Effectiveness: Reanalyze as Needed**

FIGURE 4. ANSI Z10: Hierarchy of health and safety controls (*Source:* Manuele 2005)

- elimination/minimization of the hazard through designing the facility, equipment, or process to remove the hazard, or substituting processes, equipment, materials, or other factors to lessen the hazard
- enclosing the hazard using cabs, enclosures for noisy equipment, or other means
- isolating the hazard with interlocks, machine guards, blast shields, welding curtains, or other means
- removing or redirecting the hazard with local and exhaust ventilation

2. Administrative controls. This involves changing how the employee performs workplace tasks. Examples include:

- writing operating procedures, work permits, and safe work practices
- limiting exposure times to the hazards, such as exposure to temperature extremes or ergonomic-related hazards
- monitoring highly hazardous materials used by employees
- use of alarms, signs, and warnings
- use of a buddy system
- effective use of training

3. Personal Protective Equipment. Examples include:

- respiratory protection
- hearing protection
- protective clothing
- safety glasses
- gloves
- hard hats

Protective equipment is acceptable as a hazard-control method under the following circumstances:

- when engineering controls are not feasible or do not totally eliminate the hazard
- while engineering controls are being developed
- when safe work practices do not provide sufficient additional protection
- during emergencies when engineering controls may not be feasible

The ANSI Z10 standard on occupational health and safety management systems expands upon the traditional hazard-abatement hierarchy of engineering controls, administrative controls, and personal protective equipment. The Z10 standard provides the following hazard-abatement hierarchy (ANSI/AIHA 2005, 16):

The organization shall implement and maintain a process for achieving feasible risk reduction based upon the following preferred order of controls:

A. Elimination
B. Substitution of less hazardous materials, processes, operations, or equipment
C. Engineering controls
D. Warnings
E. Administrative control
F. Personal protective equipment

Feasible application of this hierarchy of controls shall take into account:

a. The nature and extent of the risks being controlled
b. The degree of risk reduction desired
c. The requirements of applicable local, federal, and state statutes, standards, and regulations
d. Recognized best practices in industry
e. Available technology
f. Cost-effectiveness
g. Internal organization standards

This standard prescribes a hierarchy of controls that contains six elements instead of the three mandated by OSHA. The ANSI Z10 hierarchy's first priority is to design out or otherwise eliminate the hazard. If the hazard is eliminated, the overall risk associated with that specific hazard is either eliminated or reduced. Also, in ANSI Z10, the substitution element is separate from the elimination element. The additional number of elements and the separation of substitution from elimination are important changes to the prioritized hierarchy of work-site hazard-abatement activities (ANSI/AIHA 2005).

The SHE professional may use one of the control methods in the ANSI Z10 hierarchy over another control higher in the precedence when providing appropriate interim protection until the hazard is abated by a more appropriate permanent control. More likely, the selected hazard-abatement measures will be a combination of two or three items implemented simultaneously. Once the abatement measures are selected, the SHE professional should discuss these measures with all affected employees who perform the job. Employee input and feedback on hazard-control measures is important, so their responses should be carefully considered. When the hazard-control measure involves the introduction of a new or modified job procedure, the SHE professional should ensure that the employees understand the reasons for the modifications. The employees must also receive appropriate training that addresses new procedures or actions to be taken by the employee.

Training

In establishing training programs, employers must clearly define the employees to be trained and what subjects are to be covered by training. In setting up training programs, employers will need to clearly establish the goals and objectives they wish to achieve with the training that they provide to employees. The learning goals or objectives should be written in clear, measurable terms before the training begins (Reese 2003; Manuele 2003; Anton 1989; Easter, Hegney, and Taylor 2004). These goals and objectives need to be tailored to each of the specific training modules or segments. Employers should describe the important actions and conditions under which the employee will demonstrate competence or knowledge, as well as what are acceptable performance outcomes.

In 1986, OSHA issued a program evaluation profile (PEP) for their compliance officers to use when evaluating an employer's safety program (OSHA 1996). Although this compliance directive was rescinded, it serves as guidance in the evaluation of a sound employee training program. The OSHA PEP is available on the OSHA Web site. Highlights of this OSHA guidance state:

- Knowledgeable persons conduct safety and health training.
- Training is properly scheduled, assessed, and documented.
- Training covers all necessary topics and situations, and includes all persons working at the site (hourly employees, supervisors, managers, contractors, part-time and temporary employees).
- Employees participate in creating site-specific training methods and materials.
- Employees are trained to recognize inadequate responses to reported program violations.
- A retrievable record-keeping system provides for appropriate retraining, make-up training, and modifications to training as the result of evaluations.

OSHA regulations consist of more than 100 standards that contain training requirements.

OSHA has developed some voluntary training guidelines to assist employers in providing safety and health information, which are available on their Web site. These guidelines also provide employers with instructions needed for employees to work at minimal risk to themselves, to fellow employees, and to the public. A summary of the training guidelines (OSHA 2006a) lists areas designed to help employers accomplish the following:

(1) Determine whether a work-site problem can be solved by training.
(2) Determine what training, if any, is needed.
(3) Identify goals and objectives for the training.
(4) Design learning activities.
(5) Conduct training.
(6) Determine the effectiveness of the training.
(7) Revise the training program based on feedback from employees, supervisors, and other workers.

In-depth discussion of effective safety training can be found in the *Hazard Prevention Through Effective Safety and Health Training* textbook in this series.

INCIDENT INVESTIGATIONS

Thousands of workplace incidents occur throughout the United States every day. Manuele (2003, 190) defines the term *incident* as encompassing "all hazards-related events that have been referred to as accidents, mishaps, near misses, occupational illnesses, environmental spills, losses, fires, explosions, et cetera." Incident investigations are more accurately defined as a hazard-related incident. Hazards include the characteristics of things, and actions or inactions of persons that result in a potential harm. A hazard-related incident can be an unplanned, unexpected process of multiple and interacting events, deriving from the realization of uncontrolled hazards occurring in sequence or in parallel, which most likely could result in harm or damage to people, property, and/or the environment (Manuele 2003, 192).

The failure of people, equipment, supplies, or surroundings to behave or react as expected causes most of the incidents (OSHA 2005). Incidents indicate a failure by the management system of an organization, which makes it very important to investigate the root cause of an incident. Some of these reasons provided by the Indiana University of Pennsylvania Safety Department (IUP 2006, 4) include:

- Identify and correct root causes to prevent recurrence in order to reduce costs and increase profits.
- Identify trends in incident experience.
- Comply with standards and regulations.
- Satisfy public concern.
- Litigate incident claims.

Incident investigations determine how and why these management-system failures occur. By using the information gained through an investigation, a similar or perhaps more disastrous incident may be prevented. It is important to conduct incident investigations with future incident prevention in mind. The incident investigation process provides the accurate, timely information needed to prevent these recurrences (Colvin 1992).

Management's commitment to safety includes a strong visibility once incidents occur. An organization's commitment to safety can be measured by its efforts to investigate the root causes of incidents (including near misses) that cause injury to people, property, or the environment. Most occurrences of incidents are due to a failure of the management system to be able to control acts and conditions in the workplace. Once an incident occurs, management has the obligation to ensure that proper control measures are implemented to prevent them from happening again (Della-Giustina 2000). Thorough incident investigation and follow-through with remedial actions support a culture that gives importance to safety. Poor response to incident investigations give the employees reasons to doubt management's sincerity with respect to safety (Manuele 2003).

Sources of Fatal and Nonfatal Occupational Injury Statistics

The SHE professional should be familiar with national statistics on occupational injuries as part of the overall incident investigation process. National statistics on the sources of occupational injuries (fatal and nonfatal) provides the safety professional with insightful information on workplace hazards that may require evaluation. This information can be combined with internal incident investigations to identify root causes of incidents and selection of appropriate hazard abatement.

According to the National Safety Council, floors and ground surfaces were the source of 255,548 out of 1.4 million occupational injuries in 2002. Many of these injuries typically involved slips, trips, and falls with an average cost per injury in 2002 of $18,838 (Parker 2005).

Identifying the Root Cause of Incidents

The primary goal of conducting an incident investigation is to determine its root cause. Proper investigation of incidents provides information about root causes that can then be adequately corrected to prevent future occurrences. Manuele (2003, 83) defines an incident as "a term encompassing all hazards-related events that have been referred to as accidents, mishaps, near-misses, occupational illnesses, environmental spills, losses, fire, explosions, et cetera." All of the incidents Manuele defines above are derived from hazards.

The effective safety professional must break away from traditional theories of hazard reduction and accident causation and identify the root cause of the accident (Petersen 2003, 27). This point is illustrated by Petersen, using a traditional incident investigation form, in the following passage:

If we investigate a person falling off a step ladder using our present investigation forms, we identify one act and/or one condition:

> The unsafe act: Climbing a defective ladder.
> The unsafe condition: A defective ladder.
> The correction: Getting rid of the defective ladder.

Let us look at the same accident in terms of multiple causation. Under the multiple-causation theory, we would ask what were some of the contributing factors surrounding the incident:

1. Why was the defective ladder not found during normal inspections?
2. Why did the supervisor allow its use?
3. Didn't the injured employee know it should not be used?
4. Was the employee properly trained?
5. Was the employee reminded not to use the ladder?
6. Did the supervisor examine the job first?

The answers to these and other questions would lead to the following corrections:

1. An improved inspection procedure
2. Improved training
3. A better definition of responsibility
4. Pre-job planning by supervisors.

Effective safety management requires that the hazards and events that contribute to the incident process be identified, evaluated, and eliminated or controlled (Manuele 2003). There are several programs available to assist the SHE professional with performing evaluations of the root cause of an accident or incident. A few sources of these programs include:

- *Cause and Effect Diagrams.* A cause-and-effect diagram is a tool that helps identify, sort, and display possible causes of a specific problem. It graphically illustrates the relationship between a given outcome and all the factors that influence the outcome. This type of diagram is sometimes called an Ishikawa diagram because it was invented by Dr. Kaoru Ishikawa (Ishikawa 1968), or a fishbone diagram because of the way it looks. Dr. Ishikawa developed this technique to improve quality within an organization. Additional information on cause-and-effect diagrams can be found at the American Society for Quality (ASQ 2004).
- *5 Whys.* The 5 Whys is a method of solving a problem by repeatedly asking why the problem occurred, then why did that cause occur, at least five times so the layers of symptoms are peeled away until you get to the root cause of a problem. Shigeo Shingo (1981) developed this process for getting to the root cause of a problem. Very often the ostensible reason for a problem will lead to another question. Although this technique is called "5 Whys," one may find that the question will need to be asked fewer or more than five times before one finds the issue related to a problem. This process has been incorporated into the Six Sigma process of quality management. Additional information on the 5 Whys can be obtained from the iSix Sigma Web site located at www.isixsigma.com.
- *TapRooT®.* TapRooT® is a system (which includes a book, training, and software) for root-cause analysis of problems (Paradies and Unger 2000). This system assists SHE professionals in solving problems by finding and correcting root causes so that problems do not

recur. The system is based on theories of human performance and equipment performance, which are built into the TapRooT® system so that they can easily be applied. Additional information about the system is available on the TapRooT® Web site at www.taproot.com.

- *Apollo.* The Apollo process (Apollo Associated Servcies 2007) is a four-step method for facilitating a thorough incident investigation. The steps are:

  1. Define the problem. What do we want to prevent from recurring? When and where did it occur? What is the significance of the problem?
  2. Analyze cause and effect relationships. Once the problem is defined, one needs to understand the causes and how they interact with one another.
  3. Identify solutions. Solutions are specific actions that control causes.
  4. Implement the best solutions. The best solutions are those that prevent problem recurrence, are within our control, and meet our goals and objectives.

  Additional information can be obtained from the Apollo Associated Services LLC Web site at www.apollorca.com.

- *Kepner-Tregoe (KT).* Kepner-Tregoe (KT 2007) helps people consciously learn and use four basic thinking patterns to answer four basic questions:

  - What's going on?
  - Why did this happen?
  - Which course of action should we take?
  - What lies ahead?

  The Kepner-Tregoe system uses four rational processes for applying critical thinking to information, data, and experience in the process of identifying the root cause of an incident. Additional information can be obtained from their Web site at www.kepner-tregoe.com.

Investigating Incidents

Incident investigation should also be a fact-finding process, not a fault or blame-assessing process. Incidents include near-misses and accidents. Accidents result in loss, while near-misses do not (Friend 2006). As previously discussed, these incidents should be investigated to determine their root cause. The events that result in a near-miss rather than an accident may be pure chance. If our goal is to prevent future incidents, then all incidents should be investigated.

The goal of an effective safety management system should be to control or eliminate occupational hazards before there is injury or damage to people, property, or the environment. However, some hazards may not be recognized or addressed until an incident occurs. Incident investigation can be an important tool for workplace hazard identification, even though it is after the fact (Reese 2003). Identified hazards from incident investigations can be incorporated into the hazard identification and assessment process of an effective safety management system.

The investigation should be documented in writing and should adequately identify the causes of the incident and even the close-call or near-miss occurrences (ACGIH 2007).

Accident Rates Used by OSHA

OSHA tracks injury and illnesses throughout the nation. Statistics are available from OSHA on "Lost Workday Injury and Illness Rates and Total Recordable Case Rates" (BLS 2009). These rates can be computed for an organization for comparison with like organizations in a given industry. These rates are calculated as follows:

Lost-Workday Injury and Illness (LWDII) Rate

The annual LWDII rate is calculated according to the following formula, with an example and calculation provided by OSHA (2009, A-1):

$$\text{LWDII Rate} = \frac{\text{Lost-workday injuries and illnesses} \times 200{,}000}{\text{Employee hours worked}}$$

(2)

where:

- lost-workday injuries and illnesses = sum of column 2 and column 9 from the OSHA log in the reference year
- employee hours worked = sum of employee hours worked in the reference year
- 200,000 = base for 100 full-time workers working 40 hours per week, 50 weeks per year

Sample One-Year LWDII Rate Calculation

In calculating the LWDII rate of an establishment scheduled for inspection in October 2005, injury and illness cases and employment data for the preceding calendar year are used.

In this example:

- the number of LWDIIs in 2004 = 5
- the number of workers employed in 2004 = 54
- the number of employee hours worked in 2004 = 54 workers × 50 weeks = 108,000

$$\text{LWDRII Rate} = \frac{5 \times 200{,}000}{108{,}000} = \frac{1{,}000{,}000}{108{,}000}$$

$$= 9.26 \text{ (rounded to 9.3)}$$

Sample Two-Year LWDII Rate Calculation

An establishment scheduled for inspection in October 2006 employed an average of 50 workers in 2005 and 54 workers in 2004. The injury and illness cases and employment data for the two preceding calendar years are used.

The two-year LWDII rate can be calculated using the following equation.

LWDII Rate =

$$\frac{(\text{Year 1 LWDIIs} + \text{Year 2 LWDIIs}) \times 200{,}000}{\text{Yr 1 employee hrs worked} + \text{Yr 2 employee hrs worked}}$$

(3)

In this example:

- the number of LWDIIs in 2004 = 5
- the number of LWDIIs in 2005 = 6
- the number of employee hours worked in 2004 = 108,000
- the number of employee hours worked in 2005 = 100,000

$$\text{LWDRII Rate} = \frac{(5+6) \times 200{,}000}{108{,}000 + 100{,}000}$$

$$= \frac{2{,}200{,}000}{208{,}000} = 10.58 \text{ (rounded to 10.6)}$$

Three-Year LWDII Rate Calculation

When determining the rate for an employer who has been in OSHA's Safety and Health Achievement Recognition Program (SHARP) for two or more years, calculate the LWDII rate in the same way as for the two-year rate, but include the third year's data.

Total Recordable Case Rate (TRCR)

The Total Recordable Case Rate (TRCR) is the rate of total nonfatal injuries and illnesses for the calendar year reviewed. The TRCR is compared to the rate in the Total Cases column that most precisely corresponds to the Standard Industrial Classification (SIC) code of the site under review. The Total Cases column is found in the Incidence Rates table reported in the annual Bureau of Labor Statistics Data on Occupational Injuries and Illnesses.

The annual TRCR is calculated according to the following formula:

TRCR =

$$\frac{(\text{Recordable injuries} + \text{Recordable illnesses}) \times 200{,}000}{\text{Employee hours worked}}$$

(4)

where:

- recordable injuries = sum of column 2 and column 6 from the OSHA log in the reference years
- recordable illnesses = sum of column 9 and column 13 from the OSHA log in the reference year

- number of employee hours worked = sum of employee hours worked in the reference year
- 200,000 = base for 100 full-time workers working 40 hours per week, 50 weeks per year

Sample One-Year TRCR Calculation

An establishment scheduled for inspection in October 1999 employed an average of 54 workers in 1998. Therefore, injury and illness cases and employment data for the preceding calendar year are used.

In this example:

- recordable injuries = 9
- recordable illnesses = 4
- employee hours worked in 1998 = 54 workers × 50 weeks = 108,000

$$\text{TRCR} = \frac{(9+4) \times 200{,}000}{108{,}000}$$

$$= \frac{2{,}600{,}000}{108{,}000} = 24.07 \text{ (rounded to 24.1)}$$

Sample Two-Year TRCR Calculation:

An establishment scheduled for inspection in October 2000 employed an average of 50 workers in 1999 and 54 workers in 1998. The injury and illness cases and employment data for the two preceding calendar years are used.

The two-year TRCR can be calculated using the following equation.

TRCR =

$$\frac{(\text{Yr 1 Recordable data} + \text{Yr 2 Recordable data}) \times 200{,}000}{\text{Yr 1 employee hrs worked} + \text{Yr 2 employee hrs worked}}$$

(5)

In calendar year 1998:

- recordable injuries = 9
- recordable illnesses = 4
- employee hours worked = 108,000

In calendar year 1999:

- recordable injuries = 14
- recordable illnesses = 7
- employee hours worked = 100,000

$$\text{TRCR} = \frac{(9+4) + (14+7) \times 200{,}000}{108{,}000 + 100{,}000}$$

$$= \frac{6{,}800{,}000}{208{,}000} = 32.69 \text{ (rounded to 32.7)}$$

When determining the rate for an employer who has been in SHARP for two or more years, calculate the TRCR in the same way as for the two-year rate, but include the third year's data.

TOOLS OF THE TRADE

Winning Over Top Management: Obtaining a Commitment to Safety

Obtaining management leadership is crucial to the success of an effective safety management system and positive safety culture (Anton 1989, Reese 2003, Manuele 2003, Petersen 2003). Many top CEOs understand the value and savings obtained from a sound safety program, while others just perceive safety as a compliance-driven expense (ASSE 2003, Colford 2005b). Saving and cost reductions obtained through an effective safety management system has been described as a return on investment so that these economic advantages gain the attention of top management. SHE professionals have many tools available to demonstrate that an investment in safety delivers a return on an investment in safety to a company's bottom line through cost savings. However, this positive message will not be effective within an organization unless the SHE professional can truly obtain the attention and gain the interest of top management (Della-Giustina 2000).

Safety competes for top-management attention with all other departments in an organization for time and resources. The SHE professional must use salesmanship combined with the language of business to deliver the benefits and values an investment in safety generates. Selling safety can be a challenge. This is further complicated by the fact that doing the right thing should not have to be sold to management. To ease this complication, SHE professionals should not see themselves as safety salesmen, but rather as SHE professionals who use salesmanship to obtain attention, generate interest, and deliver specific benefits of safety to each audience to whom they present (Carnegie 1981).

Gaining the interest of upper management can be done in a variety of ways. One method is to illustrate the regulatory benefits of a safety management program. Manuele (2008) believes that the ANSI Z10 standard for safety management systems "will become the benchmark against which the adequacy of occupational safety and health management systems will be measured." One selling point to management is that OSHA may use the ANSI Z10 as the basis of new safety management standards that they might promulgate in the future. The American Society of Safety Engineers (ASSE) published a legal perspective of the ANSI Z10 standard (see Appendix B) that makes strong legal arguments for establishing a safety management system. This legal perspective can be used to gain the attention of upper management.

Once attention is gained by upper management the value of safety can be better demonstrated. Asking for money may generate attention (albeit negative) but may not generate enough interest to deliver the benefits of the SHE proposal. However, making a proposal that management allocate funds for safety programs, which some say will return a savings of $3 to $6 for every $1 invested (LM 2005), not only generates attention, it also generates interest.

A proposed return on investment through overall cost reductions and/or savings may generate enough interest among management to ask, "How can safety pay?" This now requires proper communication coupled with salesmanship. These facts and associated benefits presented by the SHE professional now must be put in terms to which management can relate. Selling safety based solely on reduction in injury rates, compliance with OSHA requirements, or because it is the right thing to do, may not gain management's attention when competing with other departments that present production and profit issues. Nevertheless, demonstrating that the indirect costs of a $500 emergency room visit may very well exceed $15,000 can help illustrate an accident's effect on the organization's production. This $15,000 total expenditure on an accident can be presented in terms of lost production to help gain motivation to abate the hazard. For instance, if the organization produces a product that generates a $15 net profit, it would take 1000 items to cover the cost of what appears on the surface to be a $500 injury due to the significant indirect costs associated with the incident. The SHE professional can now demonstrate that the investment in the safety management system will reduce injuries and thereby increase productivity and the company's bottom line.

Another example of how to demonstrate the savings generated by hazard-abatement activities is provided by the state of Oregon's Occupational Safety and Health (OR-OSHA) online safety training courses. OR-OSHA states, "Your supervisor may ask you what the Return on Investment (ROI) will be. If the investment to correct a hazard is $1,000, and it's likely the potential direct and indirect accident costs to the company may total $28,000 sometime in the foreseeable future (let's say five years), you can find the ROI by dividing the $28,000 by $1,000 to get 28. Now multiply that result by 100 to arrive at 2,800 percent (Total Incident Cost/Total Investment × 100 = ____% Return on Investment). Next, divide that total by 5 years to determine an annual ROI of over 500 percent. Now that's a return! Management may want to know how quickly the investment will be paid back: what the payback period is. Just divide $28,000 by 60 months

FIGURE 5. OSHA model for creating change
(*Source:* OSHA 2001a)

and you come up with $467 per month in potential accident costs. Since the investment is $1,000, it will be paid back in a little over two months. After that, the corrective action is actually saving the company some big money" (OR-OSHA n.d.).

It is important to note that research (DeArmond and Chen 2007, Chen 2005) indicates that upper management often bases opinions of safety performance, programs, and personnel on statistics (for example, workers' compensation claims/costs and injury rates). The SHE professional should also be aware of the statistics to which top-level managers give the most attention, then either emphasize those statistics in negotiations or highlight alternative data that are more reflective/indicative of actual performance (DeArmond and Chen 2007).

The final aspect of salesmanship is painting a visual picture in management's mind of the future synergy generated by a sound safety management plan (see Figure 5). The vision presented should highlight the positive effects of a sound safety management system. The positive outcomes or benefits realized to present to upper management include a large percentage reduction in worker compensation costs and insurance rates, a resulting lowered accident/injury rate and related cost savings, and the subsequent leadership by management that involves employees and generates a healthier organizational (and safety) culture. On the other hand, do not present potential outcomes that are unrealistic or unachievable.

ANSI Z10 states that, "Organizations and the community may see additional benefits of implementing an OHSMS beyond the reduction of injury and illnesses. Some of these benefits may include: lowered workers' compensation costs, reduced turnover of personnel, reduced lost workdays, compliance with laws and regulations, increased productivity, improved employee health status, improved product quality, higher morale of employees, reduction or elimination of property damage due to incidents, reduced business interruption costs, and reduced impact on the environment due to incidents" (ANSI/AIHA 2005, 6).

The benefits discussed above are the positive visions of an effective safety management system that senior management can see. The job of the SHE professional is to paint the picture of success and obtain a commitment from management while the vision is fresh. This is salesmanship. This is how to achieve commitment to safety from management.

A little salesmanship can integrate safety into the business model by illustrating incident and accident effects on production and profitability. Integrating the costs of safety into the business and demonstrating a return on investment has been identified as a major goal of the SHE professional (ASSE 2007b, ASSE/AIHA 2005). This puts safety into a language to which management, front-line supervisors, and even employees can relate. Doing the right thing is now right for many reasons.

Assessing the Culture of the Company

According to Manuele (2003), an organization's culture consists of its values, beliefs, legends, rituals, missions, goals, and performance measures, and its sense of responsibility to its employees, its customers, and its community, all of which are translated into a system of expected behavior. The culture of an organization dictates the effectiveness of a safety management system. Petersen (1996, 66) found that the culture of the organization sets the tone for everything in safety. "In a positive safety culture, it says that everything you do about safety is important." Consider this statement by OSHA: "The best Safety and Health Programs involve every level of the organization, instilling a safety culture that reduces accidents for workers and improves the bottom line for managers. When Safety and Health are part of the organization and a way of life, everyone wins" (OSHA 2002).

It is more important to understand what the culture is than to understand why it is that way (Petersen 1996). Perceptions of the safety culture of an organization can vary between management staff, various departments, front-line supervisors, and employees. The differences in perception can affect the levels of trust within an organization. A lack of trust can affect the overall performance of an organization, including production, profits, and safety attitudes. Strong management leadership combined with active employee involvement in the safety management system builds up trust and reinforces a positive safety culture.

FIGURE 6. MEMIC culture evaluation form (Source: MEMIC 2004)

Score	Safety Culture Value	Unaware 0	Beginner 1	Mediocre 2	Adequate 3	Excellent 4	Ideal 5.0
1	Continuity	Worker safety is not a factor at any time.	Safety is only applied after fines or accidents.	A safety program exists, but is not complete, and is often not applied.	A complete safety program exists, but is not always applied.	Safety is part of everything from new-hire orientations to all types of performance appraisals.	All workers at all levels understand and apply safety -- on and off the job.
2	Equality	Safety is always seen as the lowest priority.	Safety is mentioned to new hires and on workplace signs, but is rarely acted on.	Worker safety is said to be equal with production, but is not always practiced.	Safety is equal in importance to planning and budgeting, but is not always applied.	Worker safety is equally as important as planning, quality and production.	Worker safety shares an equal part in all production and performance reviews.
3	Individuality	Individual workers are not responsible for safety.	Each worker is told to be safe, but no follow-up method exists.	Sometimes individual workers take part in the safety process.	Individual workers are involved mainly through suggestion boxes or safety committees.	Individual workers are involved in all phases of the safety process.	Safety starts with each worker's personal commitment -- on and off the job.
4	Integrity	Efforts at worker safety don't exist or aren't credible.	Safety efforts come mainly from outside resources.	Safety program is administered part-time from within by workers with little training.	Some workers are assigned to safety duties full-time, but don't get regular training.	Regular training for all workers stresses that individuals apply safety to each aspect of their work.	Top management takes part in training and leads in applying the safety process.
5	Morality	Safety is done only to hold down the cost of insurance and fines.	Worker safety is talked about, but no one believes it.	Worker safety wins out over production, some of the time.	Safety training is provided to protect all workers, but not always applied.	Worker safety is a major part of planning and production at all levels.	Well-being of the worker is considered 100% of the time, including off the job.
6	Profitability	Safety is always viewed as a cost or as overhead.	Worker safety is mainly to control costs of insurance and to prevent fines.	Worker safety is said to be a priority, but is most often seen as a major cost.	Worker safety is a priority but the cost of injuries is not subtracted from profits.	The cost benefits of worker safety and the cost of injuries subtracted from profits are published.	Safety is viewed as a key asset and profit center at all levels of the company.
7	Rationality	Worker safety is not part of any planning process.	Safety is only considered after fines or accidents.	Sometimes prevention of injuries is part of the planning process.	Preventing injuries is part of the planning process, but is not always acted on.	Worker safety is a main factor in all decision making and planning.	Top management expects and plans for worker safety at all levels within the company.
8	Responsibility	No one accepts responsibility for safety.	Issues related to safety are addressed mostly by the "safety person."	Workers some of the time are responsible for safety, but usually perform better with a "safety person" around.	Workers are responsible for safety, but some of the time production wins out over safety.	Workers take action to ensure their own safety, and need limited support from a "safety person."	Every worker from top management down is responsible for safety.
9	Superiority	No efforts are made to improve any aspect of safety.	Excellence in safety is talked about, but is often not actually done.	Safety excellence is a stated priority, but no plan exists for ongoing improvement.	Excellence in safety is a stated goal, but steps to achieve this are not always taken.	Excellence in safety is achieved through budgeting for it, and by applying state-of-the-art equipment and training.	Ongoing improvement in safety is a stated goal and plans are in place to ensure this is attained.
10	Visibility	No method exists to report safety performance.	Signs or other methods report safety performance, but are not updated regularly.	Safety performance is tracked, but no goals or rewards are set.	Safety performance goals are set and made public, but are rarely attained.	Safety goals and rewards are set, and progress is published regularly companywide.	Safety performance is tracked and published at all levels equally with production and quality.

Are there any questions you would like to ask or suggestions you would like to make about this subject?

REQUIRED DATA:
○ Salaried   ○ 1st Shift   ○ Hourly
○ Hourly   ○ 2nd Shift   ○ Salaried
              ○ 3rd Shift   ○ Management

A well-written safety management program supported by management involvement sets the goals or vision of where the safety culture should be. However, it is important for the SHE professional to assess how close the current safety culture of an organization is to the desired culture. It is equally important for the SHE professional to assess the differences in the perception of the status of the safety culture between management, front-line supervisors, and employees. This perception survey can also extend among departments, facilities, regions, or any other combination of units within the organization.

Selecting the proper survey tool to assess safety culture perception is important. Reliable survey tools require careful design, implementation, and analysis. Options for conducting a perception survey include:

- What perceptions are being measured?
- Whose perceptions are being measuring?
- What survey methodology will be used?
- What questions will be asked?
- How does one test for bias or reliability of the survey?
- Who will analyze the data and how will it be reported?

Petersen (1996, 67) mentions a few considerations that determine an organization's culture:

- How decisions are made: Does the organization spend its available money on people? On safety? Or are these ignored for other things?
- How are people measured: Is safety measured as tightly as production? What is measured tightly is what is important to management.
- How people are rewarded: Is there a larger reward for productivity than for safety? This states management's real priorities.
- Is teamwork fostered? Or is it "them versus us?" In safety, is it "police vs. policed?"
- What is the history? What are the traditions?
- Who are the corporate heroes? And why?
- Is the safety system intended to save lives or to comply with regulations?
- Are supervisors required to do safety tasks daily? This says that safety is a big value.
- Do big bosses wander around? Talk to people?
- Is using the brain allowed on the work floor?
- Has the company downsized?
- Is the company profitable? Too much? Too little?

The use of an outside, independent professional can be an effective (but expensive) method for assessing the safety culture in an organization. SHE professionals who plan to conduct the culture assessment in-house should research the issue further to answer the questions posed above. Figure 6, the Safety Culture Profile Survey developed by the Maine Employers' Mutual Insurance Company (MEMIC 2004), is an example of a quick safety culture perception survey that can be used as-is or modified to fit individual organizational needs.

Culture surveys are also being researched as leading indicators of sound safety management performance. In Ontario, Canada, the Institute for Work and Health collaborated with labor and labor-based organizations to develop a set of consensus-based leading indicators that can be reliably linked to traditional lagging indicators, such as workers' compensation claims. The goal of this research is to provide leading, culture-based indicators for better managing the safety management system (Amick 2010).

The research indicated eight topic areas of an organization's culture that are considered leading indicators of safety management performance:

1. Formal safety audits at regular intervals are a normal part of our business.
2. Everyone at this organization values ongoing improvement in this organization.
3. This organization considers safety at least as important as production and quality in the way work is done.
4. Workers and supervisors have the information they need to work safely.
5. Employees are always involved in decisions affecting their health and safety.
6. Those in charge of safety have the authority to make the changes they have identified as necessary.
7. Those who act safely receive positive recognition.
8. Everyone has the tools and/or equipment they need to complete their work safely.

Once the safety culture is assessed and defined, the SHE professional can determine if the objectives of the safety management system work toward improving the culture. Metrics can be established to measure the effectiveness of the system. Areas where trust is lacking can be addressed. Poor communication and lack of participation or interest in the safety program can also be addressed. Training can be tailored to be more effective. These steps will close the perception gaps, improve trust and morale, and create a more positive safety culture within an organization.

Defining Roles for Effective Safety Management

Previous sections of this chapter established the importance of the leadership roles that must be taken by management, the involvement and active participation by employees, and oversight and program facilitation by the SHE professional. Nonetheless, a sound safety management system also assigns responsibilities, provides authority, and holds everyone accountable for the program's success. Each element of the organization's safety management system must be specifically assigned in writing to a specific job or position with coordination responsibilities and performance expectations delineated. Training should also be provided on the assigned responsibilities.

Management must grant written authority necessary to meet these assigned responsibilities (which is in the exclusive control of the individual). The employees must have confidence in their authority, understand how to exercise the authority, and then use their authority to meet assigned responsibilities in a timely manner. This authority also provides for adequate resources (personnel, methods, equipment, and funds) for the employee, along with the effective use of these resources to meet assigned responsibilities. All personnel are held accountable for meeting their safety and health responsibilities. Methods should also exist for monitoring personnel performance.

Management must have written policies for addressing situations when assigned responsibilities are not met. This can result in positive intervention, such as mentoring or coaching, or result in more negative consequences, such as verbal warnings, written warnings, demotions, or termination. On the other hand, those employees who meet or exceed responsibilities should be recognized and receive positive reinforcement for their behavior. Tracking safety performance is necessary so all responsible parties are aware of performance status and what progress has been achieved. These metrics also are used by responsible parties to adjust objectives to meet assigned safety responsibilities and continuously improve the safety management plan.

It is important to emphasize that top management should not simply delegate implementation of the safety management system to other members of the management team. Top management should remain involved and committed to the safety management system by ensuring its visible inclusion as an element of the organization's business plan (ANSI/AIHA 2005). It is also important for supervisors to be involved with safety management. Petersen (1996) found that effective supervisors received top-level safety training upon initial assignment as well as through continuing education. Effective supervisors also walked the shop floor and engaged with at least one employee a day to discuss job safety performance. Finally, Petersen found that good supervisors "make good use of the shop safety plan and regularly review job safety requirements with each worker (Petersen 1996, 294).

Manuele (2003, 76) wrote that "Employees must believe that they are responsible for their safety, and they must be provided with training, tools, and necessary authority to act." Benefits of employee involvement include:

- substantial contributions in hazard identification
- proposing solutions to hazards or problems creating hazards
- participating in solutions that reduce or eliminate the hazards

Given the necessary training and opportunity, employees can make substantial contributions to the safety program (Manuele 2003).

The Evergreen Program—Continually Improving the Process

The safety management system must be periodically reviewed. ANSI Z10, *Standard for Occupational Health and Safety Management Systems,* addresses management's review of the safety management system in this excerpt from their standard:

> "The organization shall establish and implement a process for top management to review the OHSMS at least annually, and to recommend improvements to ensure its continued suitability, adequacy, and effectiveness. Management reviews are a critical part of the continual improvement of the OHSMS" (ANSI/AIHA 2005, 3).

Subjects recommended by ANSI Z10 for at least an annual review can include (ANSI/AIHS 2005):

- progress in the reduction of risk
- effectiveness of processes to identify, assess, and prioritize risk and system deficiencies
- effectiveness in addressing underlying causes of risks and system deficiencies

This management review should also occur as systems, operations, and processes change (Anton 1989, Petersen 2003, Reese 2003). This review can include one or more of the following components:

- Evaluate the impact of new regulations and national consensus standards on the program goals and objectives.
- Examine the status of goals and objectives of the program.
- Review accident and incidents including near-miss events.
- Evaluate the scope, findings, and follow up to inspections.
- Review any training provided.
- Gauge movements in the safety culture of the organization, including both management and labor.
- Evaluate observable employee safe work-practice behaviors.
- Evaluate the physical conditions of the workplace.

Manuele (2008) found several typical deficiencies or provisions of a safety management system where shortcomings exist. These include:

- risk assessment and prioritization
- application of a prescribed hierarchy of controls to achieve acceptable risk levels
- safety design reviews
- inclusion of safety requirements in procurement and contracting papers
- management of change systems

This review can be conducted by an individual (or team) determined competent in all applicable areas by virtue of education, experience, and/or examination. The results of the review are documented and drive appropriate changes or adjustments in the program.

Previously identified deficiencies do not appear on subsequent reviews as deficiencies.

A process should exist that allows deficiencies in the program to become immediately apparent and corrected, in addition to requiring a periodic, comprehensive review. There must be a demonstration that the safety management system results in the reduction or elimination of hazards, incidents, and accidents (Manuele 2003).

Defining the Professional Safety Practitioner

Robert DeSiervo offers this broad definition of a safety professional (DeSiervo 2004).

> Safety Professionals are individuals who are engaged in the prevention of events that harm people, property or the environment. Occupational safety professionals help organizations prevent injuries, illnesses and property damage. These professionals must acquire knowledge of safety sciences through education and experience so that others can rely on their judgment and recommendations. They use qualitative and quantitative analysis of simple and complex products, systems, operations, and other activities to identify hazards. They evaluate the hazards to identify what events can occur and the likelihood of occurrence, severity of results, risks (a combination of probability and severity), and cost. They identify what controls are appropriate and their cost and effectiveness. Safety Professionals make recommendations to managers, designers, employers, government agencies and others. Hazard controls may involve administrative controls (such as plans, policies, procedures, training, etc.) and/or engineering controls (such as safety features and systems, fail-safe features, barriers and other forms of protection). Safety Professionals may manage and provide help to implement controls.

Both conventional wisdom and a review of the literature reveal an ever-increasing need to produce

SHE educators and SHE practitioners in this country in order to meet the challenges confronting the SHE profession in the twenty-first century (DeLeo 2003).

The demand for competent and qualified SHE professionals continues to increase each year (ASSE 2007b). However, defining what makes a SHE professional qualified and competent often engenders a difficult and controversial discussion. Occupational safety and health management in the United States and throughout the world is, and always has been, dynamic. Therefore, those who work in the safety, health, and environmental professions must also be dynamic. One size does not fit all. There are some SHE competencies that stretch across most, if not all, professions, such as regulatory record keeping, ergonomics, hazard communications training, personal protective equipment, and job hazard-analysis concepts. Other professions require industry-specific SHE knowledge and skill sets. This includes industries in the healthcare, transportation, mining, construction, and environmental arenas. These industry-specific professions require the knowledge, competencies, and experience to address a much more defined set of regulatory compliance and job-specific occupational hazards and issues.

Safety is everyone's responsibility. But assigning safety responsibilities does not guarantee the success of a safety program. The importance of effective safety awareness and training as a component to developing a strong and effective safety culture in an organization is critical, as previously mentioned. At what point does an individual require higher levels of training, a more formal safety education or degree, on-the-job experience, and skill sets in safety to succeed? Job descriptions and responsibilities can identify the safety knowledge, education, and skills necessary to be competent in SHE-related duties. Entry-level positions may or may not require any formalized safety education or experience. Many employees find safety has been added to their current job descriptions as safety programs are implemented and expanded.

Consider the following positions that interface with an effective safety management system. What education, experience, and certifications make these individuals competent in their safety responsibilities?

- a part-time fire/safety officer (also employed as a local fire fighter) who checks fire extinguishers monthly, performs new employee fire safety training, and performs limited fire safety inspections throughout the facility
- an operations manager at a manufacturing plant who is responsible for employee safety
- a laboratory manager who is responsible for the chemical hygiene plan and OSHA compliance issues, in addition to chemical waste management, disposal, and emergency response
- a personnel manager who has the dual responsibility for the overall safety program
- a safety technician working within a larger safety department who conducts general safety awareness training and departmental safety inspections and issues PPE
- a union mechanic who is elected as the safety representative on the company safety committee
- a corporate safety director for a Fortune 500 company who is responsible for safety management for worldwide operations
- a CFO who understands how safety can reduce insurance costs and improve the bottom line
- a CEO who recognizes that good safety management improves morale and production of the workforce
- shareholders who benefit from improved productivity and reduced costs from a strong safety management program

Each of these job descriptions interfaces with the overall safety management system to different degrees. The uniqueness of SHE issues of a given industry and/or job description requires a variety of competencies and experience levels that must be adapted and molded for overall SHE management to be effective. So what makes a SHE professional competent? What experience, education, certifications, and skills are necessary for the competent safety practitioner to succeed?

Defining competencies can be a moving target. What competencies are needed for today's safety practitioner? What curricula or competencies may best prepare students to become the competent safety managers of tomorrow (Blair 2003)? What competencies are required of academia to provide curriculums that serve both of these groups? Competencies for the safety practitioner are defined by experience, skills, certifications, education, and an ever-changing body of knowledge.

Experience

Education and certification aside, there is no substitute for on-the-job experience in effective safety management. Years of on-the-job safety experience can develop skills that can be difficult to duplicate or measure through education or certification alone. Those who have witnessed near misses, accidents, injuries, and fatalities gain unique experiences that can change an individual forever. These first-hand experiences have been converted into learning tools to shock employees into learning the consequences of unsafe behavior. OSHA created a document called *Fatal Facts* (OSHA 2009) that captures unsafe behaviors and relates them to real-life occupational fatalities.

The experience that qualifies a competent safety practitioner is defined as *supervised experience*. Supervision by other experienced, competent safety professionals may be especially valuable to recent entrants into the field (DeSiervo 2004). Supervised experience can bring about anticipation of potential hazards by recognizing unsafe activities or conditions. However, workplace experience, such as experience shared by line workers, can also provide great knowledge and insight for the new safety practitioner. An experienced line foreman might know more than a Certified Safety Professional (BCSP) about the safe operation of the equipment he operates, including lockout/tagout, machine guarding, and personal protection use.

Workplace experience can sometimes be a negative. Repetitive workplace experience can lull those who know better into complacency and carelessness when performing repetitive tasks. Workplace experience can also teach us tricks to bypass safety devices designed to protect employees and equipment. Safety practitioners should seek a balance between supervised experience and workplace experience.

Experience has its financial benefits, too. Safety practitioners with ten to twenty years' experience earn about $18,000 more per year than those with less than ten years' experience. Figure 7 breaks down average salary by years of experience:

Unfortunately, training and experience alone can be difficult to quantify. Many certifications and designations were designed to help measure an individual's competency in a given area. Finding a certification or educational opportunity that specifically fits the unique occupational requirements the safety practitioner needs to satisfy may prove to be difficult. It may be necessary for a safety practitioner to demonstrate or measure his or her competency in safety in order to advance and succeed. This can only be accomplished through the pursuit of academic degrees and/or professional certifications.

Education

"Safety is a relatively new academic discipline" (Adams et al. 2004, 1). Prior to the passage of the Occupational Safety and Health (OSH) Act of 1970 (OSHA 1970), there were few college or university safety courses or degree programs available for those entering the safety profession (ASSE 2003). Safety in academia emerged during the 1970s and has blossomed into a legitimate educational path at the associate, baccalaureate, masters, and doctorate levels of higher education. The American Society of Safety Engineers provides the following overview for higher education opportunities for the safety practitioner:

> A number of community and junior colleges offer an associate degree in safety or a related field. People

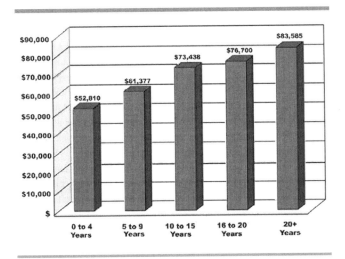

FIGURE 7. Average income for SHE professionals (*Source:* ASSE 2004)

graduating from these programs are hired for limited positions in safety. They may help manufacturers, construction companies or other industries meet OSHA's hazard control standards. A number of four-year colleges and universities offer undergraduate degrees in safety. Today, over 90% of those in the safety profession have earned at least a bachelor's degree. About 30% of those entering the field have a bachelor's degree in safety, while many move into safety from other disciplines (engineering, business, physical sciences, etc.) and later pursue safety studies. About 40% of today's safety professionals have advanced degrees. Some of those with an advanced degree in safety graduated with a bachelor's degree in a non-safety field. They may use a master's degree in safety to prepare for and enter the safety profession. Some who get their safety preparation at the bachelor's level also pursue graduate study in safety or a safety-related specialty, such as industrial hygiene, environmental science, public health or ergonomics. Some work toward advanced degrees in related fields, such as business and engineering that will enhance their career opportunities. (ASSE/BCSP 2007a)

As of May 2005, there were over 70 safety, health, and environmental-related programs offered by community colleges, colleges, and universities. Nearly 20 percent of these courses were Accreditation Board for Engineering and Technology (ABET)-accredited (ASSE 2007b). There are also many educational institutions offering SHE-related degrees, including environmental, industrial hygiene, engineering, and fire science programs. Many of these programs are available through distance learning programs that allow the SHE professional to obtain educational credentials over the Internet or through the mail. These emerging forms of distance learning provide a more convenient, self-paced opportunity for those who do not have accessible SHE programs available to meet individual needs.

An education provides a strong foundation for those entering the safety profession. It also can accelerate promotions and compensation. A study published by the American Society of Safety Engineers found that the average income for safety practitioners by education ranged from about $60,000/year for those with a high school education, to over $75,000/year for those with a bachelor's degree, to nearly $95,000/year for those with a Ph.D or Ed.D (ASSE 2004). Acquiring a formal degree in safety will also aid individuals in obtaining professional recognition, certifications, and credentials. Many safety certifications accept SHE academic achievements as prerequisite qualifications. Other professional safety organizations, including ASSE, BSCP, and AIHA, recognize and use academic qualifications to reduce experience prerequisites or to take the place of exams.

Accredited Education Program Criteria

According to *The Career Guide to the Safety Profession* (ASSE/BCSP 2007a, 19),

> To prepare for the safety professional courses, college students are normally required to take courses in mathematics through beginning calculus, statistics, chemistry with laboratory work, physics with laboratory work, human physiology or biology, and introductory courses in business management, engineering mechanics and processes, speech, composition, and psychology. Students in safety must also acquire good computer skills, including the ability to use the Internet and important business and safety software packages. Professional safety courses in higher education include safety and health system management, design of engineering hazard controls, industrial hygiene and toxicology, fire protection, ergonomics, environmental safety and health, system safety, accident/incident investigation, product safety, construction safety, educational and training methods, assessment of safety performance, and behavioral aspects of safety. Students may also elect to take specialty courses beyond the required courses.

In 1996, ABET embarked on an accreditation reform effort designed to foster an environment in which each graduate of engineering, technology, computing, and applied science, where the safety curricula reside, possesses the skills necessary for both lifelong learning and productive contributions to the profession, employers, economy, and society (ABET 2010). Accreditation criteria have changed from subject matter content to outcome-based activity (ASSE 2003). The major outcome criteria include a baseline understanding of math and science; analysis and interpretation of data; anticipating, identifying, and evaluating hazards; developing hazard control design, methods, procedures, and programs; the functioning of multidisciplinary teams and group activities; understanding SHE professional responsibilities; a knowledge of contemporary SHE issues in both global and societal contexts; communication skills; and training skills.

The ASSE Educational Standards Committee (ESC) developed revised safety curriculum guidelines in early 2005 to assist universities in the development of safety curricula and also to assist ASSE Program Evaluators participating in the accreditation process (ASSE 2005a). The revised criteria became effective during the 2006 evaluation cycle. The criteria that ABET Baccalaureate Safety Programs include must demonstrate that graduates are able to:

1. Anticipate, recognize, and evaluate hazardous conditions and practices affecting people, property, and the environment.
2. Develop and evaluate appropriate strategies designed to mitigate risk.
3. Apply principles of safety and health in a nonacademic setting through an internship, cooperative, or supervised experience.

"Topics for Safety Curriculum" identifies the appropriate topics to include in safety curricula (ASSE 2005a). Table 1 identifies both required topic areas as well as recommended areas. It is important to remember the topic areas are broad areas that were not intended to be all-inclusive. For example, within the Safety Management Topic, a program may choose to cover a variety of topics such as human behavior, cost-benefit analysis, organizational and management theory, safety management systems, safety program elements, record keeping, and so on. Where possible, all of these topic areas should be tied in with the three major program objectives discussed at the beginning of these guidelines that deal with recognition, evaluation, and control. Universities are encouraged to contact the ASSE Educational Standards Committee for further assistance in the development of safety curricula.

The American Society of Safety Engineers identified four purposes that institutional accreditation serves (ASSE 2005a):

1. It assures a level of quality.
2. It is required for student access to certain federal funds, such as student aid.
3. It eases transfer. Many regionally accredited colleges and universities will only accept transfer credit or admission to graduate school from students from regionally accredited colleges or universities.
4. It engenders employer confidence.

Current or future safety practitioners should evaluate educational opportunities carefully. Accredited programs offer a valuable third-party assurance of the curriculum of a given program.

Certifications

The question has been posed (Blair 2003), "Do safety, health, and environmental practitioners need to be Certified Safety Professionals?" What other certifications might benefit the safety practitioner? There are over 300 different certifications worldwide that one can achieve in the disciplines of safety, health, industrial hygiene, and hazardous wastes. Over 300 SHE designations, with only a handful meeting widely accepted accreditation requirements, create a confusing, even dangerous, practitioner landscape (ASSE 2005b). Many of these certifications are accredited by multiple third parties requiring rigorous qualifications, testing, and maintenance. Other certifications merely require an individual to fill out an application and pay an appropriate fee. Obtaining safety certification does not necessarily equate to safety competence. Therefore, what is a legitimate certification for the competent SHE professional? Gaining the right certification for one's industry can improve job qualifications, responsibilities, and compensation.

Certifications have demonstrated a positive impact on the average compensation paid to safety practitioners (ASSE 2007b, BCSP 2009, AIHA 2008). A compensation survey by the American Society of Safety Engineers (ASSE 2004) found that those who have some type of extra certification or credential earned on average $12,500 more than those who do not have any certification. One of the most recognized accredited safety credentials in the world is the Certified Safety Professional (CSP) designation administered by the Board of Certified Safety Professionals (BCSP). BSCP found that the average income for CSPs was

> **SIDEBAR 6**
>
> **NEW JERSEY SAFETY PROFESSIONAL TRUTH IN ADVERTISING ACT**
> **(State of New Jersey 2002)**
>
> New Jersey P.L.2002, c.050 (S894 1R) CHAPTER 50
> An Act concerning safety professionals and supplementing P.L.1960, c.39 (C.56:8-1 et seq.).
>
> Be It Enacted by the Senate and General Assembly of the State of New Jersey:
>
> C.56:8-113 Short title.
>
> 1. This act shall be known and may be cited as the "Safety Professional Truth in Advertising Act."
>
> C.56:8-114 Findings, declarations relative to qualification of safety professionals.
>
> 2. The Legislature finds and declares that it is necessary to provide assurance to the public that individuals holding any safety certification have met certain qualifications.
>
> C.56:8-115 Definitions relative to qualifications of safety professionals.
>
> 3. As used in this act:
>
> "Safety profession" means the science and art concerned with the preservation of human and material resources through the systematic application of principles drawn from such disciplines as engineering, education, psychology, physiology, enforcement and management for anticipating, identifying and evaluating hazardous conditions and practices; developing hazard control designs, methods, procedures and programs; implementing, administering and advising others on hazard controls and hazard control programs; and measuring, auditing and evaluating the effectiveness of hazard controls and hazard control programs.
>
> A "Safety professional certification organization" means a professional organization of safety professionals which has been in existence for at least five years and which has been established to improve the practice and educational standards of the safety profession by certifying individuals who meet its education, experience and examination requirements. The organization shall be accredited by the National Commission of Certifying Agencies (NCCA) or the Council of Engineering and Scientific Specialty Boards (CESB), or a nationally recognized accrediting body which uses certification criteria equal to or greater than that of the NCCA or CESB.
>
> C.56:8-116 Certification by safety professional certification organization required.
>
> 4. It shall be an unlawful practice for any person to advertise or hold himself out as possessing a professional safety certification from a safety professional certification organization unless that person is certified by the applicable safety professional certification organization.
>
> 5. This act shall take effect on the first day of the 25th month following enactment.
>
> Approved August 3, 2002.

$99,448. Another highly recognized safety certification is the Certified Industrial Hygienist (CIH) designation administered by the American Industrial Hygiene Association (AIHA). A 2008 AIHA salary survey also confirmed the value of safety certifications to compensation (AIHA 2008). Their study found that safety practitioners with a CIH received average compensation of $10,340. Compensation for both of those certifications is well above the $79,809 average compensation AIHA found for those with no certification at all.

In today's economic and business climate, the safety practitioner is not likely to work for the same employer his or her whole career. Brauer and Murphy (2005) found that third-party safety certifications can also provide lateral mobility to safety practitioners by providing those holding a quality, recognized certification with credentials to compete for positions with other employers (Hintch 2005).

Certifications that have been accredited by a recognized third-party accreditation body are supported by many professional safety organizations and government agencies. Because the work of safety professionals has a direct impact on public safety and health, government organizations, employers, and those awarding contracts are concerned that safety professionals be fully qualified and competent to do their jobs. Safety professionals may therefore need other credentials in addition to their education degree. These credentials might include licenses, registration, and professional certification. To date, no state requires safety professionals to be licensed in order to practice (ASSE 2005b).

Third-party safety-accredited certifications have been incorporated into regulations, professional qualifications, and position statements for title protection of safety practitioners. One state has taken the lead in defining what constitutes a safety professional. New Jersey has enacted legislation to ensure those claiming the title of "safety professional" meet minimal standards. The "Safety Professional Truth in Advertising Act" is reproduced in part in Sidebar 6.

Many OSHA regulations cite accredited safety certifications for reviewing and/or modifying existing safe work practices at job sites. Safety associations, including the American Society of Safety Engineers, American Industrial Hygiene Association, National Safety Council, and the Institute of Hazardous Materials Managers, require third-party accredited certifications as the preferred way to achieve professional status in their respective organizations (see "Appendix A: Recommended Reading," for a list of organizations and their Web sites). Widely accepted organizations that provide highly reliable, quality mechanisms for accrediting certifications include:

1. The National Commission on Certifying Agencies (NCCA), established originally by the federal government to help ensure capabilities of federal allied health practitioners recognized under federal law
2. The Council of Engineering and Scientific Specialty Boards (CESB)
3. The Organization for International Standardization (ISO) under the American National Standards Institute

These are the only accreditation organizations that meet legal and psychometric standards closely similar to those followed by state licensing mechanisms (ASSE 2005a).

Many safety associations and certifying bodies have also pushed for title protection for the safety practitioners they represent. AIHA actively promotes and defends the CIH designation in a variety of legislative venues on local, state, federal, and international levels. These efforts are focused and effective since all of their professional members are CIHs. However, there are other, more diverse safety groups that have difficulty representing and satisfying a majority of their members with such diverse backgrounds. The American Society of Safety Engineers (ASSE) represents over 30,000 safety, health, and environmental professionals throughout the world. Only about 25 percent of its membership has achieved the status of CSP, CIH, or CHMM (ASSE 2005b). ASSE fully understands that there are uncertified members with experience, education, and training who perform as competently as those with these accredited certifications. However, to advance the profession as a whole through the regulatory and legislative process, ASSE has the responsibility to rely on the highest level of demonstrated competence to achieve that goal. The ASSE has developed a position paper for proposed safety regulations stating:

> When legislators and regulators insert a profession in a bill or regulation, they typically look to an accepted third party to provide some measure of the professional's qualifications. ASSE is not aware of a legislator or regulator who would be willing to write in an SHE practitioner simply because of their years of employment, education achievements, or unique experiences without those qualifications being validated by a widely accepted third party. States typically are not willing to provide such certification programs. Without such third-party validation of qualifications, every piece of regulation or legislation would be required to include a unique set of practitioner qualifications as well as a process to ensure the appropriateness or validity of those qualifications. (ASSE 2005b).

When selecting a certification, it is important to evaluate the many values it can provide, including (BCSP 2009):

- any third-party accreditation
- the incorporated body of knowledge
- prerequisites, qualifications, and necessary experience
- professional recognition
- professional growth and compensation
- regulatory recognition

The benefits of obtaining third-party accredited safety certification have demonstrated a positive impact on the professional development, advancement, and compensation of safety practitioners.

Competencies and Skills

Soule (1993, 6) found safety practitioners need to be more than just compliance technicians—"they must be good communicators who understand business language and possess good management skills." In his 2003 study, Dr. Earl Blair stated, "Knowledge of business, accounting, and marketing are topics that can meaningfully expand the safety, health and environmental practitioners should they be integrated into the safety curriculum. Safety professionals may lose credibility, if although otherwise highly competent in the technical aspects of safety, they appear ignorant in these important areas" (Blair 2003). Business skills are a must for the successful safety practitioner. They need to possess the total package in order to be successful (DeSiervo 2004).

According to the Blair study (2003), the business skills that have been identified as needing further definition include:

1. The particular areas of business expertise where safety managers need to be most knowledgeable.
2. The specific topics and curriculum most useful for promoting effective communication and interpersonal skills.
3. The most effective ways to measure safety performance that incorporate a broad spectrum of measures, such as trailing indicators, cost indicators, and leading indicators.

Additional competencies were identified and recommended for those seeking to develop doctoral-level safety programs for teaching future SHE professionals. William DeLeo conducted a study asking each member of the two groups, safety educators and practitioners/experts, to submit and rank safety competencies (DeLeo 2003). Those competencies that received a 100 percent ranking in the study's finding included:

1. Ability to effectively use the safety sciences literature.
2. Ability to critically evaluate existing research literature in occupational safety and health, and identify gaps in that literature.
3. Developing, implementing, managing, and evaluating safety and health programs.
4. Evaluating safety and health program performance and designing performance measures.
5. Ability to develop hazard control methods, procedures, and programs to include integration of safety performance into the goals, operations, and productivity of organizations and their management.
6. Ability to estimate the economic impact of safety issues and practices on a firm's economic performance (cost modeling).

Those competencies that received a 93 percent importance ranking in the study included:

1. Demonstrated in-depth knowledge of safety, health, and environmental issues.
2. Ability to stay current with the changing safety challenges in the workplace.
3. Ability to review, compile, analyze, and interpret data from accident and loss event reports and other sources regarding injuries, illnesses, property damage, environmental effects, or public impacts in order to identify causes, trends, and relationships.
4. Ability to understand and articulate what is an appropriate safety and health management system for an organization in an industry with a specific culture at a certain point on the life cycle of a business.

Recent safety literature, professional development conferences, and seminars have been inundated with multiple topics on safety business skills, including leadership, safety culture, business of safety, and management. These articles, books, research papers, and presentations provide the safety practitioner with an abundance of data to help fill in any gaps in the skills and knowledge necessary to succeed. These current competencies must be achieved to succeed in today's business climate. These competencies must be monitored and reevaluated to ensure our current safety students are prepared to succeed in the future as well.

The Safety Body of Knowledge (BoK)

The American Society of Safety Engineers defines the safety body of knowledge references of the SHE profession as ". . . a collection of commonly used general and specific information on the safety, health, and environmental theory, principles, and practice in the broad field of loss prevention and control" (ASSE 2003).

While the definition of the body of knowledge of the SHE profession has not yet been consolidated into one document, the ASSE has generally defined one as including the following elements:

- standards used to accredit SHE curriculum for colleges and universities
- ASSE-published documents and materials used to describe the scope and functions of the professional SHE position
- requirements for certification or licensure as an SHE professional, as currently stated by the Board of Certified Safety Professionals
- the ASSE Code of Professional Conduct
- SHE regulations, standards, and other legislation that define SHE compliance
- publications, books, and materials used by SHE professionals to implement effective SHE management programs

ASSE prepared a white paper defining a Safety Body of Knowledge in June 2003. One of the most significant findings published in the white paper was that safety practitioners use reference codes, standards, and regulations more than any text, and also use the Web extensively (ASSE 2003). However, there is strong consensus among safety leaders that significant portions of our body of knowledge continue to evolve. Therefore, the safety profession as a whole must continue to address the issue of an ever-changing body of knowledge reference for the safety, health, and environmental profession that identifies those resources needed for the prevention of injury, illness, death, and the destruction of property and the environment.

Competencies for the Future Safety Practitioner

The American Industrial Hygiene Association also looks to the future of the safety practitioner in an expanding global arena by stating:

> . . . AIHA believes all parties (government, employers, and employees) should begin to discuss the changing workplace and expansion of the global economy and assume a lead role in offering recommendations to address occupational health and safety in the global economy of the 21st century. (AIHA 2004)

A study by DeLeo (2003) identified academic criteria necessary to meet this future need when it stated:

> New training programs and techniques must be designed to meet the needs and challenges of the 21st century workplace. One of the resultant competencies exiting the final round of this study was an ability to stay current with the changing safety challenges in the workplace.

Safety certification bodies are also looking to adjust to the future needs of safety practitioners. The BCSP has identified the need to study the flexibility in qualifications and examinations in its strategic plan. They are looking to expand the path to the safety profession for those who enter the profession with an adjunct role or as technicians and technologists (Brauer and Murphy 2005).

Competency requirements for the future safety practitioner will address a balance between supervised experience, higher education, and third-party certification. The competent safety practitioner will combine these competencies with character traits guided by professional ethics and a professional code of conduct to create a trustworthy safety practitioner. Both character and competency will be necessary for the safety practitioner to meet the challenges present in the twenty-first century.

REFERENCES

Abrams, A. 2005. Email communication to author (May 16).

———. 2006. *Legal Perspectives of ANSI Z-10* (retrieved July 31, 2010). www.asse.org/membership/docs/92ArticleaboutZ10LegalPerspectives.pdf

Accreditation Board for Engineering and Technology (ABET). 2010. *History* (retrieved August 31, 2010). www.abet.org/history.shtml

Adams, P., R. Brauer, T. Bresnahan et al. 2004. "Professional Certification." *Professional Safety* 49(12): 26–31.

American Conference of Governmental Industrial Hygienists (ACGIH). 2007. *How to Develop a Simple, Cost-Effective Safety & Health Program: A Small Business Guide* (retrieved March 31, 2007). www.acgih.org/about/committees/SBguide.htm

American Industrial Hygiene Association (AIHA). 2008. "Compensation & Benefits Survey." Fairfax, VA: AIHA.

American National Standards Institute (ANSI) and American Industrial Hygiene Association (AIHA). 2005. *ANSI/AIHA Standard Z10-2005: Occupational Health and Safety Management Systems*. Fairfax, VA: AIHA.

American Society for Quality (ASQ). 2004. *Quality Tools: Fishbone Diagram* (retrieved March 24, 2007). www.asq.org/learn-about-quality/cause-analysis-tools/overview/fishbone.html

American Society of Safety Engineers (ASSE) and Board of Certified Safety Professionals (BCSP). 2002. *White Paper Addressing the Return on Investment for Safety, Health, and Environmental (SHE) Management Programs* (retrieved March 31, 2007). www.asse.org/bosc_articles_2.htm

———. 2007a. *ABET/ASSE Accredited Safety Degree Programs* (retrieved December 11, 2007). www.asse.org/professionalaffairs/education/directory/directory_abet.htm

———. 2007b. *Career Guide to the Safety Professional*. 3d ed. Des Plaines, IL: ASSE and BCSP.

American Society of Safety Engineers (ASSE). 2003. *White Paper of the Body of Knowledge Task Force of the American Society of Safety Engineers Council on Practices and Standards* (retrieved February 4, 2008). www.asse.org/practicespecialties/bok/docs/bok_wpapers6-03.pdf

———. 2004. "ASSE Compensation Survey." *Professional Safety* (October) 49(10):26–27.

———. 2005a. *Safety Curriculum Guidelines* (retrieved February 4, 2008). www.asse.org/professionalaffairs/govtaffairs/ngpost19.phpvarSearch=Safety+Curriculum+Guidelines

———. 2005b. *ASSE Position Statement on ASSE Government Affairs Representation of SHE Professionals*. Des Plaines, IL: American Society of Safety Engineers.

Amick, B. C. 2010. "Managing Prevention with Leading and Lagging Indicators in the Workers' Compensation System" in *Use of Workers' Compensation Data for Occupational Injury & Illness Prevention* (pp. 83–87). Washington, D.C.: Department of Health and Human Services.

Anton, Thomas J. 1989. *Occupational Safety and Health Management*. New York: McGraw Hill.

Apollo Associated Services. 2007. *The Apollo Process* (retrieved December 13, 2007). www.apollorca.com/process/process.shtml

Barfield, G. 2004. "Safety in Business Terms." *Professional Safety* 49:8.

Blair, E. H. 1999. "Which Competencies Are Most Important for Safety Managers?" *Professional Safety* (October) pp. 28–32.

———. 2003. "Culture and Leadership." *Professional Safety* (June) pp. 18–22.

Board of Certified Safety Professionals (BCSP). 2009. *CSP Facts* (retrieved July 31, 2011). www.bcsp.org/Salary_Survey

Brauer, R., and H. Murphy. 2005. "Workplace Safety Delivers ROI." *Human Capital Magazine* (June).

Bureau of Labor Statistics (BLS). 2006. *Industry Injury and Illness Data* (retrieved December 13, 2007). www.bls.gov/iif/oshsum.htm

Carnegie, Dale. 1981. *How to Win Friends and Influence People*. Rev. ed. New York: Simon and Schuster.

Chen, P. Y. 2005. How Can We Increase Safety Behaviors? Speech to the Chemistry Division at Los Alamos National Laboratory, Los Alamos, New Mexico.

Colford, J. 2005a. "CEOs Who Get It." *Safety and Health* (February) 171(2):25–33.

———. 2005b. "The ROI of Safety: It Starts at the Top." *Business Week* (September).

Colvin, R. J. 1992. *The Guidebook to Successful Safety Programming*. Chelsea: Lewis Publishers.

DeArmond, S., and P. Y. Chen. 2007. Financial Executives' Perceptions of Safety Performance, Safety Programs, and Safety Personnel. Paper presented at the XIIIth European Congress of Work and Organizational Psychology, May 9–12, Stockholm, Sweden.

DeLeo, W. 2003. "Safety Educators and Practitioners Identify the Competencies of an Occupational Safety and Environmental Health Doctoral Degree: An On Line Application of the Delphi Technique." *Journal of SH&E Research* (Spring) 1(1).

Della-Giustina, D. E. 2000. *Developing a Safety and Health Program*. Boca Raton, FL: CRC Press.

DeSiervo, Robert. 2004. "The Education of a Safety Professional." *Journal of SHE Research* (Fall) 4(2).

Downs, D. E. 2003. "Management System Assessment." *Professional Safety* 48(11):31–38.

Easter, K., R. Hegney, and G. Taylor. 2004. *Enhancing Occupational Safety and Health*. Boston: Elservier.

Friend, M., and J. P. Kohn. 2006. *Fundamentals of Occupational Safety and Health*. New York: Government Institutes.

Hintch, B. 2005. "Interview with Roger Brauer." *Compliance Magazine* (May) pp. 3–6.

Indiana University of Pennsylvania. 2006. *Accident Investigation and Analysis* (retrieved December 13, 2007). www.coned.iup.edu/SafetyScience/OCCSafety/os2.htm

International Labour Office (ILO). 2001. *Guidelines on Occupational Safety and Health Management Systems*. Geneva, Switzerland: ILO.

International Occupational Hygiene Association (IOHA). 1998. *Occupational Health and Safety Management Systems*. Geneva, Switzerland: IOHA.

Ishikawa, Kaoru. 1968. *Guide to Quality Control*. Tokyo, Japan: Asian Productivity Organization.

Kepner-Tregoe, Inc. 2007. *The KT Way* (retrieved December 13, 2007). www.kepnertregoe.com/TheKTWay/OurProcesses.cfm

Liberty Mutual Research Institute for Safety (LM). 2005. *Liberty Mutual Workplace Safety Index*. Hopkinton, Massachusetts: Liberty Mutual Research Institute for Safety.

Maine Employers' Mutual Insurance Company. 2004. *Safety Culture Profile Survey*.

Manuele, F. A. 2003. *On the Practice of Safety*. 3d ed. Hoboken, NJ: John Wiley & Sons.

_____. 2005. "Risk Assessment & Hierarchies of Control." *Professional Safety* 50(5):33–39.

_____. 2008. *Advanced Safety Management*. Hoboken, NJ: John Wiley & Sons.

National Institute for Occupational Safety and Health (NIOSH). 2005. *Publication No 2005-112: A Compendium of NIOSH Economic Research 2002–2003* (retrieved February 11, 2008). www.cdc.gov/niosh/docs/2005-112/default.htm

Occupational Safety and Health Administration (OSHA). 1970. *OSH Act of 1970* (retrieved February 15, 2008). www.osha.gov/pls/oshaweb/owasrch.search_form?p_doc_type=OSHACT

_____. 1989. *Safety and Health Program Management Guidelines; Issuance of Voluntary Guidelines; Notice*. Federal Register Vol. 54, No. 16, Jan. 26, 1989, pp. 3904–3916.

_____. 1996. *Program Evaluation Profile (PEP)* (retrieved August 31, 2010). www.osha.gov/SLTC/safetyhealth/dsg/topics/safetyhealth/pep.html

_____. 1998a. *Draft Proposed Safety and Health Program Rule. 29 CFR 1900.1. Docket S&H-0027* (retrieved February 4, 2008). www.osha.gov/dsg/topics/safety-health/nhsp.html

_____. 1998b. *OSHA 2254, Training Requirements in OSHA Standards and Training Guidelines (revised)*. Washington, D.C.: OSHA.

_____. 1998c. *OSHA Software Expert Advisors, OSHA's $afety Pays E-tool* (retrieved August 31, 2010). www.osha.gov/dcsp/smallbusiness/safetypays.html

_____. 2001a. *Safety and Health Management Systems E-tool* (retrieved August 31, 2010). www.osha.gov/SLTC/etools/safetyhealth/index.html

_____. 2001b. *Directive 00-01 (CSP 02), TED 3.6, Consultation Policies and Procedures Manual, Chapter 1, Section IX: A Brief History of the OSHA Consultation Program* (August 6, 2001) (accessed February 28, 2008). www.osha.gov/pls/oshaweb/owadisp.show_document?p_table=DIRECTIVES&p_id=2584#8-I

_____. 2002. *OSHA 3071, Job Hazard Analysis (revised)*. www.osha.gov/Publications/osha2254.pdf

_____. 2004. *Voluntary Protection Plans, Recognizing Excellence in Safety* (retrieved August 31, 2010). www.osha.gov/Publications/vpp/vpp_kit.pdf

_____. 2005. *OSHA 2209-02R, Small Business Handbook*. www/osha.gov/Publications/smallbusiness/small-business.pdf

_____. 2006a. *Accident Investigation* (retrieved August 31, 2010). www.osha.gov/SLTC/accidentinvestigation/index.html

_____. 2006b. *06-06 (CSP 02)—TED 3.6, Consultation Policies and Procedures Manual, Chapter 8: OSHA's Safety and Health Achievement Recognition Program (SHARP) and pre-SHARP* (retrieved March 17, 2007). www.osha.gov/pls/oshaweb/owadisp.show_document?p_table=DIRECTIVES&p_id=3489

_____. 2007. *Fatal Facts Accident Reports* (retrieved March 10, 2007). www.osha.gov/OshDoc/toc_FatalFacts.html

_____. 2009. *Fact Sheet, Voluntary Protection Programs*. www.osha.gov/OshDoc/data_General_Facts/factsheet-vpp.pdf

Oregon OSHA (OR-OSHA). 2006. *Online Course 100, Safety and Health Management Basics* (retrieved August 31, 2010). www.cbs.state.or.us/Esternal/osha/educate/training/pages/100outline.html

Paradies, M., and L. Unger. 2000. *TapRooT, The System for Root Cause Analysis, Problem Investigation, and Proactive Improvement*. Knoxville, TN: Systems Improvement, Inc.

Parker, J. G. 2005. "Stopping Injuries Means Getting Down to the Source." *Safety and Health* (May) pp. 6–8.

Petersen, Dan. 1996. *Human Error Reduction and Safety Management*. Hoboken, NJ: John Wiley and Sons.

_____. 2003. *Techniques of Safety Management: A Systems Approach*. Des Plaines, IL: American Society of Safety Engineers.

_____. 2004. "Leadership & Safety Excellence: A Positive Culture Drives Performance." *Professional Safety*. 49(10), 28–32.

Reese, C. D. 2003. *Occupational Health and Safety Management*. Boca Raton, FL: Lewis Publishers.

Schneid, T. 2000. *Modern Safety and Resource Control Management*. New York: John Wiley & Sons.

Shingo, S. 1981. *A Study of the Toyota Production System.* Tokyo, Japan: Productivity Press.

Soule, R. 1993. Perceptions of an Occupational Safety Curriculum by Graduates, Their Employers and Their Faculty. PhD diss, University of Pittsburgh.

State of New Jersey. 2002. *Safety Professional Truth in Advertising Act.* New Jersey P.L. 2002, c.050 (S894 1R) Chapter 50.

Walton, M. 1986. *The Deming Management Method.* New York: Perigree.

Workplace Safety and Insurance Board and Canadian Manufacturers and Exporters Ontario Division. 2001. *Business Results Through Health and Safety* (retrieved August 31, 2010). www.wsib.or.ca/wsib/wsibsite.nsf/LookupFiles/DownloadableFileBusinessResultsThroughHealth&Safety/$File/Biz.pdf

APPENDIX A: RECOMMENDED READING

Codes, Regulations, and Standards

American National Standards Institute (ANSI) standards.

Department of Transportation. 49 CFR. *Transportation.*

Environmental Protection Agency. 40 CFR. *Protection of the Environment.*

National Fire Protection Association (NFPA) codes and standards.

Occupational Health and Safety Administration. 1974. 29 CFR 1910. *Occupational Safety and Health Standards.* Washington, D.C.: U.S. Department of Labor

Occupational Health and Safety Administration. 1992. 29 CFR 1926. *Safety and Health Regulations for Construction.* Washington, D.C.: U.S. Department of Labor.

Texts, References, and Research Publications

Alli, B. O. 2001. *Fundamental Principles of Occupational Health and Safety.* Geneva, Switzerland: International Labour Office.

Deming, W. E. 1994. *The New Economics for Industry, Government, Education.* Cambridge, MA: MIT Center for Advanced Education Studies.

Downs, D. E. 2003. "Management System Assessment." *Professional Safety* 48(11):31–38.

Muller, S., and C. Braun. 1998. *Safety Culture—A Reflection on Risk Awareness.* Zurich, Switzerland: Swiss Reinsurance Company.

National Institute for Occupational Safety and Health (NIOSH). 2004. Publication #2004-135. *How to Evaluate Safety and Health Changes in the Workplace.* Cincinnati, OH: NIOSH.

National Safety Council (NSC). 2006. *Accident Prevention Manual.* Itasca, IL: NSC.

———. 2006. *Supervisors Safety Manual.* Itasca, IL: NSC.

———. 2007. *Injury Facts.* Itasca, IL: NSC.

Noncommercial Web Sites

American Industrial Hygiene Association (www.aiha.org)

American National Standards Institute (www.ansi.org)

American Society of Safety Engineers (www.asse.org)

Bureau of Labor Statistics (www.bls.gov)

Centers for Disease Control and Prevention (www.cdc.gov)

National Fire Protection Association (www.nfpa.org)

National Institute for Occupational Safety and Health (www.niosh.gov)

National Safety Council (www.nsc.org)

U.S. Department of Labor (www.dol.gov)

U.S. Department of Transportation (www.dot.gov)

U.S. Environmental Protection Agency (www.epa.gov)

U.S. Occupational Safety and Health Administration (www.osha.gov)

Journals and Periodicals

Industrial Safety and Hygiene News. BNP Media.

Journal of Occupational and Environmental Hygiene. American Industrial Hygiene Association. (www.aiha.org/Content/AccessInfo/joeh/)

NFPA Journal. National Fire Protection Association. (www.nfpa.org/journalPortal.asp?categoryID=187&src=NFPAJournal)

Occupational Hazards Magazine. Penton Media. (www.occupationalhazards.com/)

Occupational Health and Safety Magazine. Stevens Publications. (www.ohsonline.com/)

Occupational Safety & Health Reporter. Bureau of National Affairs. (www.bna.com/products/ens/oshr.htm)

Professional Safety Journal. American Society of Safety Engineers. (www.asse.org/professionalsafety/)

Safety + Health Magazine. National Safety Council. (www.nsc.org/shnews/)

The Synergist. American Industrial Hygiene Association. (www.aiha.org/Content/AccessInfo/synergist/)

Workplace HR & Safety. Douglas Publications, LLC. (www.workplacemagazine.com/)

Classic Texts (initial publication date of 1974 or earlier)

Carson, Rachel. 1962. *Silent Spring.* Boston: Houghton Mifflin; Cambridge, MA: Riverside Press.

Clayton, George D., and Florence E. Clayton, eds. *Patty's Industrial Hygiene and Toxicology.* 1st ed. New York: Wiley.

Hammer, Willie. 1975. *Occupational Safety Management and Engineering*. Englewood Cliffs, NJ: Prentice-Hall.

Heinrich, H. W. 1931. *Industrial Accident Prevention, a Scientific Approach*. New York: McGraw Hill Book Company.

National Safety Council. 1946. *Accident Prevention Manual for Industrial Operations*. Chicago, IL: Wm. H. Pool Co.

———. 1970. *Fundamentals of Industrial Hygiene*. Chicago, IL: National Safety Council.

Petersen, Dan. 1971. *Techniques of Safety Management*. New York: McGraw-Hill.

Simonds, Rollin, and John Grimaldi. 1956. *Safety Management. Accident Cost and Control*. Homewood, IL: R.D. Irwin.

APPENDIX B: ANSI Z15.1 STANDARD: A TOOL FOR PREVENTING MOTOR VEHICLE INJURIES AND MINIMIZING LEGAL LIABILITY

By Adele Abrams, Esq. (Copyright © by Adele Abrams LLC)

Motor vehicle crashes that occur on American roadways have historically been the leading cause of occupational fatalities in this country. In the decade between 1992 and 2001, more than 13,000 civilian workers died in such incidents—accounting for 22 percent of all injury-related deaths. According to the Occupational Safety and Health Administration (OSHA), every 12 minutes someone dies in a motor vehicle crash, every 10 seconds an injury occurs and every 5 seconds a crash occurs.[1] Moreover, despite overall decreases in the number and rates of occupational fatalities from all causes, the annual number of work-related roadway deaths has actually increased to a rate of 1.2 deaths per 100,000 full-time employees.[2] The majority of such crash victims are male (89 percent), and the toll is highest among 35–54 year old workers (47 percent).

Although, as expected, persons employed in the transportation industry make up the predominant occupational sector involved in motor vehicle crashes, other affected sectors include the service industry (14 percent), manufacturing (8 percent), and sales (7 percent). What is significant from a legal perspective is that 62 percent of the vehicles occupied by a fatally injured worker were registered to a business or to the government; 17 percent were driver-registered, and just 12 percent were registered to an entity or individual that was not connected to the driver.[3]

Employers whose workers are involved in such crashes have tremendous liability exposure, especially if the individuals injured or killed are third parties (non-employees), where no worker's compensation liability shield exists as an exclusive legal remedy. They bear not only the worker's compensation costs for their employees, and the potential damage awards from third-party tort claims, but also the costs of equipment replacement and the indirect costs of workforce disruption and lost productivity associated with such incidents.

Motor vehicle crashes cost employers $60 billion annually in medical care, legal expenses, property damage, and lost productivity. OSHA estimates that the average crash costs an employer $16,500. When a worker has an on-the-job crash that results in an injury, the cost to their employer is $74,000. Costs can exceed $500,000 when a fatality is involved.[4] If punitive damages are awarded, that figure can soar into the millions of dollars per incident.

The actions of drivers employed by a company, including their failure to inspect the motor vehicle for defects as well as any unsafe behaviors while driving a company vehicle, can be imputed to the employer

---

[1] See http://www.osha.gov/Publications/motor_vehicle_guide.html.

[2] Centers for Disease Control, *Roadway Crashes Are the Leading Cause of Occupational Fatalities in the U.S.*, DHHS (NIOSH) Publication No. 2004-137 (March 2004).

[3] Census of Fatal Occupational Injuries (CFOI), 1992-2001, Bureau of Labor Statistics, and Fatality Analysis Reporting System (FARS), 1997-2002, National High Traffic Safety Administration (NHTSA).

[4] See NHTSA, *The economic burden of traffic crashes on employers: costs by state and industry and by alcohol and restraint use*. Publication DOT HS 809 682 (2003).

under the legal theory of *respondeat superior*.[5] Under this analysis, in the event of a work-related accident on a public roadway, all a tort attorney will need to demonstrate in order to name the employer as defendant in a personal injury or wrongful death lawsuit is that the company exercised some degree of control over the driver, and that the accident occurred while the driver was acting in the course of the employment relationship. Each state will apply its own twist to the vicarious liability doctrine.[6]

Other legal fault doctrines that can apply to employers arising from occupational motor vehicle incidents include:

- Negligent hiring/retention (failure to exercise due care when hiring workers who will drive in the course of their activities by checking driving records etc.)[7];
- Negligent supervision (failing to take corrective action where the employer becomes aware of prior incidents, tendencies toward aggressive or distracted driving);
- Negligent training (failure to provide appropriate documented training for the type of vehicle that the worker will operate); and
- Owner liability (failure to ensure that its agents inspect the vehicles appropriately to prevent operation with known defects, or negligent entrustment of the owner's vehicle to an unqualified or impaired individual).[8]

Even in those situations where the employer's own workers are the only victims of roadway incidents, there may be exclusions if the employer is found to be grossly negligent, as certain states permit tort actions to go forward in such circumstances or enhance the monetary awards available under worker's compensation programs.

Thus prevention through development of proactive initiatives is critical to preserve life, property and to avoid incurring the monetary costs associated with occupational motor vehicle incidents. The Liberty Mutual Insurance Company reported in 2001 that 61 percent of surveyed business executives believe their companies receive a Return on Investment of $3.00 or more for every $1.00 they spent on improving workplace safety.[9] In the case of occupational motor vehicle incidents, the underlying causes of these fatalities and injuries vary widely from mechanical failure to poor highway and vehicle design to driver error. Preventive measures also vary widely, including preventive vehicle maintenance, increased seat belt use, effective driver training, anti-lock brakes, road maintenance and safer vehicle design.

The causes and solutions are so varied that there is no single, simple strategy for prevention. There is, however, a new tool that can be utilized by employers, consultants, insurance industry experts, and other safety and health professionals to help reduce the occupational casualties, high costs, and legal liability associated with motor vehicle incidents. The American

---

[5]*Respondeat superior*, or "vicarious liability," is a key doctrine in the law of agency, which provides that a principal (employer) is responsible for the actions of his/her/its agent (employee) in the "course of employment. By definition, motor vehicle accidents that occur while a worker is in the course of his employer's business (whether or not operating an employer-owned vehicle) would fall within the scope of this legal theory, and the driver's negligence would be imputed to the employer for purposes of litigation.
[6]For example, under Maryland law, courts focus on whether the incident arose from employees' activities within the scope of the employment. To satisfy the legal test, the conduct must be of the kind the employee is employed to perform and must occur during a period not unreasonably disconnected from the authorized period of employment in a locality not unreasonably distant from the authorized area, and actuated at least in part by a purpose to serve the employer. See *Jordan v. Western Distributing Company*, 135 Fed.Appx. 582 (4th Cir. 2005). The conduct must also be expectable or foreseeable. *Sawyer v. Humphries*, 587 A.2d 467, 471 (Md. 1991).
[7]This cause of action focuses on the employer's negligence in selecting the individual as an employee, rather than on the employee's wrongful act itself. See *Van Horne v. Muller*, 705 N.E.2d 898 (Ill. 1998). In a negligent selection claim, there normally is a rebuttable presumption that an employer uses due care in hiring an employee. See, e.g., *Evans v. Morsell*, 395 A.2d 480, 483 (Md. 1978).
[8]However, if the employer is a governmental entity, sovereign immunity may apply.
[9]See Liberty Mutual Insurance Company, *Liberty Mutual Executive Survey of Workplace Safety* (2001).

Society of Safety Engineers has released the ANSI Z15.1-2006 national consensus standard, *Safe Practices for Motor Vehicle Operations*.

The standard, approved by ANSI on February 15, 2006, took effect on April 28, 2006. It provides guidelines and establishes best practices for development of motor vehicle safety programs for all classes of employers—whether addressing a single vehicle or a fleet, whether the equipment is employer-owned, employee-owned, or leased from a third party.[10] It includes such key components as:

- Management, leadership and administration;
- Operational environment;
- Driver considerations;
- Vehicle considerations; and
- Incident reporting and analysis.

As noted in the ANSI Z.15.1, when developing a program to control risks associated with motor vehicle operation, it is critical to include both operator training and qualification criteria as well as a system for inspecting and maintaining the equipment. Although inspections are normally conducted in a systematic way by drivers who have commercial driver's licenses and operate large trucks that require CDL compliance, this step is too often ignored for passenger vehicles or for smaller trucks that may be used by sales and service personnel.

The Z15.1 standard includes these components, as well as methodologies for record keeping, reporting of motor vehicle-related incidents and data/trend analysis that can be used to prevent recurrences. This is a particularly significant component from a legal perspective, as employers who are found to have actual knowledge of program failures or unsafe actions/conditions and who fail to take appropriate remedial action are much more likely to be found grossly negligent in the event of a subsequent incident. This can, of course, lead to high dollar OSHA penalties,[11] as well as punitive damages in the tort law arena arising from personal injury or wrongful death suits. In particularly egregious circumstances, there could even be criminal prosecutions targeting management personnel who were aware of deficiencies and failed to take appropriate corrective action.

Among the critical features of the ANSI Z15.1 standard are attention to driver error and the risk factors arising from driver impairment and distraction as well as the high-profile issue of aggressive driving practices, which is being criminalized in some states. The standard also emphasizes safety considerations when purchasing or modifying motor vehicles.

The standard could be used as an affirmative defense during litigation. As a recognized national voluntary consensus standard (benchmark), an employer potentially could use the standard as an indicator that it implemented programs to enhance safety for its motor vehicle operations. Use of the standard, and the ability to document compliance with the standard, could also be used as an affirmative defense when contesting federal and state citations.

In the 1990s, motor vehicle safety was designated as one of the Occupational Safety and Health Administration's priority issue areas.[12] In July 1990, OSHA issued a Notice of Proposed Rulemaking for a standard, which would have required seat belt use and driver awareness programs.[13] Although this rulemaking effort was stalled, in part due to congressional

---

[10] The standard is not intended to apply to off-road equipment, agricultural equipment, recreational vehicles, haul trucks operated solely on industrial or mine sites, or unlicensed equipment.

[11] In addition to utilizing the ANSI Z-15.1 standard as a resource in program development, employers should also be aware of OSHA Guidelines for Employers to Reduce Motor Vehicle Crashes, which can be found at: http://www.osha.gov/Publications/motor_vehicle_guide.html.

[12] Today, OSHA continues to focus on this subject through its Alliances, including those with ASSE, the Independent Electrical Contractors, the Air Conditioning Contractors of American, the Network of Employers for Traffic Safety and the National Safety Council.

[13] The Occupational Safety and Health Administration's proposed rule was published at 55 FR 28728 (July 12, 1990). That rule contained a mandatory safety belt requirement applicable to anyone driving or occupying any motor vehicle that is company owned, leased or rented or privately owned when used for official business on public highways and off highway. In addition, it included a driver training requirement for workers operate motor vehicles for official business on highway and off highway.

action that urged OSHA to further study the issue before proceeding, the agency can still regulate this recognized threat to safety through Section 5(a)(1) of the OSH Act, the "General Duty Clause."[14]

It should be noted that the standard does include this language in the Foreword section of the standard: *This standard is not intended to serve as a guide to governmental authorities having jurisdiction over subjects within the scope of the Z15 Accredited Standards Committee (ASC).* But even absent formal rulemaking, ANSI Z15 serves as a valuable reference. It also could have possible enforcement ramifications under the General Duty Clause (discussed above) by federal OSHA. It may be employed to satisfy regulatory requirements of certain state-plan OSHA programs. A number of States have enacted laws mandating such traffic management programs for employers,[15] so adoption of ANSI Z15 potentially at the state level may satisfy the compliance obligations for employers in those jurisdictions. Insurance companies encourage their client companies to implement safety and health management programs, and therefore utilization of Z15 potentially could generate monetary savings on insurance (both liability and worker's compensation).

The OSH Act covers every employer engaged in a business affecting interstate commerce who has one or more employees. By contrast, the Secretary of Transportation, acting through the Office of Motor Carrier Safety (OMCS), exercises statutory authority over the operation of motor vehicles engaged in interstate or foreign commerce.[16] However, the Department of Transportation still defers to OSHA to enforce safety related to motor vehicles where the Federal Motor Carrier Safety Administration standards in Title 49 of the Code of Federal Regulations do not address particular safety issues. Thus, reduction of work-related motor vehicle accidents is properly part of OSHA's 2003–2008 Strategic Management Plan. Of course, the U.S. Department of Transportation reserves the authority to regulate "commercial motor vehicles" which include, among others, vehicles with a gross vehicle weight rating of 10,001 pounds.[17]

Moreover, the National Advisory Committee on Occupational Safety and Health has recommended that OSHA promulgate a standard addressing motor vehicle safety, and that it involve other governmental agencies as well as safety organizations. Under OMB Circular A-119, which requires that any federal government agency rulemaking consider extant consensus standards and adopt those standards where feasible, the ANSI Z15.1 standard could eventually be incorporated by reference into a future OSHA rulemaking on this issue. OSHA also has a memorandum of understanding with ANSI (1/19/2001). The memorandum notes that: *ANSI and OSHA will maintain a mechanism for consultation in the planning of occupational safety and health standards development activities in the areas of mutual concern to the extent consistent with OSHA policy and section 6 of the OSH Act;*

---

[14]Section 5(a)(1) of the Occupational Safety and Health Act of 1970 (OSH Act) requires employers to "furnish to each of his employees employment and a place of employment which are free from recognized hazards that are causing or are likely to cause death or serious physical harm to his employees."

[15]See, e.g., Cal-OSHA's standard at http://www.dir.ca.gov/title8/8406.html.

[16]OMCS authority is found in title 49 of the United States Code in the following sections: 3101 et seq.; 2301 et seq. (known popularly as the Surface Transportation Assistance Act); 1801 et seq., dealing with the transportation of hazardous materials; and 2501 et seq. (known popularly as the Motor Carrier Safety Act of 1984).

[17]See, e.g., 49 USC §31132. Another distinguishing factor is that the term "employer" under the Motor Carriers Safety Act of 1984 means ". . . any person engaged in a business affecting commerce who owns or leases a commercial motor vehicle in connection with that business, or assigns employees to operate it," but such term does not include Federal, State, and local governments. Thus, in the case of the term "employer" under the Motor Vehicle Safety Act, there is a limitation on the OMCS jurisdiction. If, in any factual circumstances involving a section 4(b) (1) controversy between OSHA and the OMCS, where the employer does not come within the Motor Carrier Safety Act's definition of the term "employer," OSHA would have jurisdiction over the employer's working conditions and could enforce unsafe conditions or actions imputable to the employer under the General Duty Clause.

The OMB Circular (consistent with Section 12(d) of the National Technology Transfer Assistance Act (NTTAA)) directs agencies to use national consensus standards in lieu of developing government-unique standards, except when such use would be inconsistent with law or otherwise impractical. However, under the current OSH Act, only national consensus standards that have been adopted as, or incorporated by reference into, an OSHA standard pursuant to Section 6 of the OSH Act provide a means of compliance with Section 5(a)(2) of the Occupational Safety and Health Act, 29 U.S.C. § 651 et seq. ("the OSH Act").[18] Therefore, at some future time, Z15 could be adopted by OSHA as a mandatory safety and health standard through notice-and-comment rulemaking.

Another significant area of possibility would be development of consent orders with government agencies involving motor vehicle operations. There is the possibility of the standard being used as a benchmark for an employer to use in establishing such programs. The use of voluntary national consensus standards to settle such cases is a common practice and there is the possibility of Z15 being used in such a manner.

From a defensive strategy, employers who adhere to the recommendations in ANSI Z15.1 will not only see a reduction in the motor vehicle incidence rate but will also have appropriate documentation, such as written motor vehicle safety programs, safety policies, and maintenance programs and records, to reduce the likelihood litigation in the first instance because of the due diligence provided to elimination of motor vehicle risk factors. Application of ANSI Z15.1's recommendations concerning driver recruitment, selection and assessment, orientation and training, and impaired/distracted/aggressive driver prevention programs, can also be useful in defeating claims of negligent recruitment and retention of employees that might otherwise arise in third-party injury actions.

Finally, the attention to regulatory compliance and management program audits will help minimize the potential for enforcement actions brought by OSHA under the General Duty Clause, relevant DOT agencies, or even state and local governmental agencies under traffic and criminal laws.

Finally, ANSI Z15 has possible value in constructing settlement agreements or consent orders with federal OSHA, state-plan OSHA agencies or other state and federal transportation-related agencies. Often employers who have systemic safety problems will be encouraged or required, as a condition of abatement or settlement, to design and implement programs that will address management failures in a cohesive manner. The scope and function of Z15 would likely satisfy the enforcement goals of prevention of future safety issues while encouraging penalty reductions to offset the costs of program implementation. There is the strong potential of the standard being included in settlement proceedings for occupational safety and health citations involving motor vehicle operations.

SH&E professionals should be encouraged to take the following actions:

- Obtain a copy of this standard, review the standard and the background materials about it, and discuss it with senior management and legal counsel so that all parties are aware of what is expected. A legal opinion written by corporate counsel would also be a prudent action to take.
- Write and publish a policy addressing Z15 in regard to how it fits in with the organization's current program and the U.S. Occupational Safety and Health Act and the rules and regulations of the U.S. Department of Transportation. Write, implement, and document communication structures detailing how information is passed up the communication chain to senior management.
- Conduct through assessments to identify significant SH&E exposures and the means used

---

[18]Specific national consensus standards [e.g., American National Standards (ANSI) standards], which the Secretary of Labor adopted on May 29, 1971, were either used as a source standard and published in Part 1910 as an OSHA standard or explicitly incorporated by reference in an OSHA standard.

to communicate them to those in a position of authority.
- The Z15 Standard potentially could place accountability on senior management. There is some correlation with the requirements of Sarbanes Oxley Act of 2002 Public Law 107-204. It is important to ensure that SH&E audits are independent and that the results are reported and acted upon. Those ES&H practitioners who author/sign those audit reports and who fail to follow-up on the recommended actions may be subject to sanctions such as listed under the new law. The point has been made that they now have a duty that goes beyond just informing management.
- Follow the ASSE Code of Conduct.

In summary, ANSI Z15 provides safety and health professionals with a significant new tool to help enhance existing program design or to help smaller employers create a program that can protect workers while at the same time satisfying regulatory entities and insurers, effectuating cost savings and minimizing legal liability.

Managing a Safety Engineering Project

3

Joel M. Haight

LEARNING OBJECTIVES

- Recognize and understand the relevant concepts of managing project work.

- Be able to analyze project status to ascertain budget.

- Establish a schedule for the overall performance elements of project work.

- Understand the relevant concepts of leadership, team building and interaction, and managing conflict.

- Be able to analyze one's own leadership abilities.

- Be able to analyze the interpersonal and managerial skills of others.

To PROPERLY TREAT the subject of managing safety engineering work, some discussion should be provided about what safety engineering is. It is difficult to define since it is one specialty in the larger and more broadly defined safety and health discipline, and, as such, it is often loosely defined and discussed.

Paul Wright in his book *Introduction to Engineering* (1989) states that the ABET, Inc. (formerly known as the Accreditation Board of Engineering Technology) defines *engineering* as "A profession in which a knowledge of mathematical and natural sciences gained by study, experience and practice is applied with judgment to develop ways to utilize, economically, the materials and forces of nature for the benefit of mankind."

ABET defines an *engineer* as a person, who by reason of their special knowledge and use of mathematical, physical, and engineering sciences and the principles and methods of engineering analysis and design, acquired by education and experience, is qualified to practice engineering (Duderstadt, Knoll, and Springer 1982). Wright (1989) adds to this definition: "The engineer's knowledge must be tempered with judgment. Solutions to engineering problems must often satisfy conflicting objectives and the preferred optimum solution does not always result from a clean-cut application of principles or formulas." He also states that "Engineers are concerned with the creation of structures, devices and systems for human use."

In the foreword of James CoVan's book *Safety Engineering* (1995), Rodney D. Stewart writes: "Safety Engineering is an increasingly important and growing 'horizontal' dimension of engineering that cuts across all the traditional vertical dimensions (civil, mechanical, electrical, chemical and software)." CoVan writes: "Safety is a broad, multidisciplinary topic and this book addresses the engineering aspects of the subject." This indicates that there are engineering aspects of safety that must be managed.

Safety engineering is a diverse and often poorly understood subject. Many critics doubt that engineering is involved. "A weakness of the discipline that developed to serve the underlying need is that many of its practitioners were untrained and undisciplined in its application" (CoVan 1995). Roger Brauer writes in his book *Safety and Health for Engineers* (2005): "Safety engineering is devoted to application of scientific and engineering principles and methods to the elimination of hazards. Safety engineers need to know a lot about many different engineering fields." Gloss and Wardle (1984) write in their book *Introduction to Safety Engineering*: "Safety engineering is slowly maturing as a recognized profession. Safety engineering is a relatively new profession and reflects our mounting concerns for the environment, the consumer and the rights of workers." This lends importance to the need to apply established management practices to this new field.

Many aspects of the safety engineer's job fit the ABET definition for engineering. While safety engineers do not often create "structures" or "devices" as noted by Wright (1989), it can be strongly argued that the safety engineer does create "systems" for the benefit of humankind.

This chapter does not attempt to formally define safety engineering work, but proposes a working definition for the purposes of discussing specific management concepts, principles, and activities associated with projects and work in which safety engineering is inherent. These management concepts, principles, and activities include scheduling, manpower and other resource allocation, budgeting effectiveness, measurement systems, purchasing, work definition, and so on, associated with projects involving the design, construction, installation, operation, maintenance, and dismantling and/or disposal of equipment, systems, processes, or facilities in which safety engineering is an integral component. These types of projects may include excavation protection systems, fall-protection systems, energy-isolation systems, confined-space-entry systems, general construction, and general maintenance turnaround activities.

It is important to recognize that safety is no different than any other aspect of a project, or of work in general. It is often presented and discussed as a separate entity that must be accomplished and that must be treated individually. However, for the purposes of this chapter, *safety* is not a noun describing a specific activity; in most cases throughout the chapter, it will be used as an adjective describing the way all work gets done—*safely*. Even though it is not a separate entity, the safety-related aspects of the job have to be managed just like all other aspects of any project. For example, budgeting, developing specifications, ordering adequate quantities, and dispensing respiratory protective equipment must be done for a confined-space-entry job just as these functions must be done for the structural support members for a bridge, where fastening devices would be necessary equipment rather than respiratory protectors.

This chapter provides discussion on what a project is, as well as what a managed system is. It discusses where safety engineering fits into a project or system. It covers general management principles such as organizing, defining the work, scoping, scheduling, budgeting, and staffing. Workforce issues such as training and learning, motivating, team building, conflict, and leadership are covered, and the chapter also addresses work and workforce analysis concepts such as performance ratings, work sampling, allowances, time study, and resource allocation. Final products and deliverables are discussed, and many of the concepts are further illustrated in the form of examples and open-ended problems to work out.

INTRODUCTION TO MANAGING SAFETY ENGINEERING WORK

A *system* is defined throughout the literature in many ways, but for the purposes of this book, consider a system to be any process involving the interaction between humans and equipment in which raw materials (input) are converted to final products (output) (Eisner 2002). Managing the safety engineering aspects of work requires that one focus on the interaction between humans and equipment as well as on the conversion process of raw materials to final products.

This interaction involves such issues as workers responding, in the form of physical action, to signals from equipment in the process of its operation. The interaction could also involve workers being exposed to hazards associated with the raw materials or intermediate products or to hazards inherent in the equipment itself. The conversion process will likely involve moving parts, conveyor-belt operations, changing chemical states, heating or cooling, changing structural conditions, changing pressures, and so on. The manufacture of paper, pharmaceuticals, beer, or televisions, the refining of oil, the production of electricity, all involve work done by systems. The safety engineering aspects of these systems must be managed.

A *project* can be described as a formal gathering of people and equipment in an industrial setting, working toward satisfying a set of goals, objectives, and requirements. A project usually has a defined time period, a limited scope, and an established budget, and is managed by a project manager. In many cases, a project deals only with one aspect of a system's life cycle, such as design, construction, installation, or maintenance. Projects have a safety engineering aspect just as a system does, and many times this aspect is related to exposure to hazards, time pressure to complete the work, or implementation with people who may not yet be trained or experienced (Eisner 2002). The safety engineering aspects of a project must therefore also be managed.

What does it mean to manage safety engineering work associated with projects or systems? To answer this question, one must go back to the traditional key elements of managing any type of work. The classic definition of management usually includes a discussion of planning, organizing, leading, and controlling the operation for the purpose of productivity (Schermerhorn 1993).

Managers rely on a number of different management approaches to control their operations. A *classical approach* to management includes scientific management, administrative principles, and bureaucratic organizations. It assumes all people are rational. The *scientific approach* to management is one that Fredrick Taylor helped to develop in 1911. His principles involve developing a "science" for every job that would include the rules of motion, standardized work tools and proper working conditions, carefully selecting the right person for each job, properly training each worker for the job, giving them the right incentives, and supporting the workers by planning their work (Freivalds and Niebel 2008). The *behavioral approach* to management assumes that people are social and self-actualizing. It assumes people act on the basis of desires for satisfying social relationships, responsiveness to group pressures, and the search for professional fulfillment. In the *quantitative approach* to management, managers focus on the use of mathematical techniques for managing the problem solving of the operation. *Modern approaches* to management focus on the total system and look at the business as one interrelated big picture, utilizing contingency thinking and an awareness of a more global picture. Each approach has its own merits, advantages, and disadvantages, and one should understand the system under which he or she is operating to allow for a smooth integration.

PROJECT PLAN
Objectives

Objectives are usually statements that express the desires and expectations of the organization managing an overall operation. Objectives for the safety engineering aspects of the project often are developed as a result of a safety, health, fire, or compliance problem and are identified through the various analytical and evaluation methods used in the safety engineering community. Once a safety-related problem is identified, objectives for its resolution must be developed. These objectives describe expected accomplishments that will help to resolve or correct the safety-related problem and ensure the success of the organization. Usually, when safety engineering work results from a problem, the scope is bigger than just managing the implementation of corrective actions. Correcting the problem may require implementing a large-scale, resource-demanding effort. Such an effort must be cost effectively and efficiently managed. Examples of project-level objectives might include: ensure all operations personnel learn to operate the waste water plant to ensure human-error-induced catastrophic incidents do not occur, reduce exposure to noise in the compressor operation area to a level that does not exceed 50 percent of the allowable daily dose (ADD), or reduce the risk of a catastrophic release of chlorine from the five one-ton storage cylinders in the cooling water operation to a value of lower or tolerable risk. To achieve these objectives will likely require the implementation of a large-scale, project-driven course of action.

Project Requirements

Once the objectives are defined, it is the project coordinator who must determine the next course of action. This cannot take place without first identifying all of the *requirements* that must be met to safely and efficiently achieve the stated objectives. This will have a large bearing on nearly all aspects of the project, namely, schedule, cost, satisfaction of objectives, satisfaction of regulatory demands, resource allocation, and so on. While each project will have its own specific requirements, objectives, audience, and scope, there are some standard areas that must be addressed by the project safety engineer that can be generalized across all projects. Considering project requirements from a generalized approach will help analysts be sure they have addressed every necessary requirement and worked them into the project plan (Eisner 2002).

Customer Objectives: Technical and Functional

One must start with the requirements of the customer. What objective must be achieved to satisfy the person or organization requesting the accomplishment? In the field of safety engineering, the customer is often an operation's management team, and, to a large extent, the workforce. The customer's objective is likely to be a safer operation, a product made with less risk of incident, fewer injuries, fewer failures, less damage, and so on. These objectives are general in nature, so the safety engineer in charge of managing the project must be sure that agreement is reached or approval given for the means by which the objective will be achieved. For example, if the objective is to reduce benzene exposures plantwide, a project team must assemble to determine and agree on what achieving the objective will entail. Will it be achieved through the use of personal protective equipment and training? Or will engineering controls and hardware installation be required, such as zero-leak valves, enclosed and ventilated sample stations, and activated-carbon filtration systems on building air intakes?

The customer's requirements will usually be technical and have functional requirements such as fire resistance, structural integrity, spacing and layout, level of system automation and control-system logic, warmth, adequate illumination, vessels that meet appropriate pressure ratings (see the Pressure Vessel chapter in the Risk Assessment and Hazard Control section of this handbook), accessibility (foot, reach, and traffic), ease of maintenance, or personal protective equipment (PPE) comfort and aesthetics.

An additional factor that must be considered by the project manager and discussed among the project team is the potential for future expansion and growth. This must be considered from a space and real-estate point of view as well as from the standpoint of interface and integration of old and new systems. In other words, it is important to leave a system owner with the ability to tie old and new together without having to worry about complete replacement due to obsolescence of interfaces, fittings, power requirements, control-system logic, and so on.

Regulatory Objectives

Once the customer's requirements are determined, the project manager (or, in the case of a safety-related project, the safety engineer) must determine the requirements of government agencies, local municipalities, and consensus organizations. Regulatory requirements that must be considered include those involving proximity to local businesses and residences, process and storm water runoff (during construction and normal operations), odors, airborne concentrations, noise levels (during construction and normal operations), local building and fire codes, evacuation and emergency vehicle access/egress, rail or trucking access, pressure vessel or boiler codes, and Americans with Disabilities Act (ADA) access requirements. A project manager will meet and work with agency representatives (or internal compliance representatives) to identify all of the applicable regulatory requirements that must be satisfied and then estimate the time needed to satisfy the demands of all the agencies so that the project schedule includes this time. Sometimes project managers consider only the technical and functional requirements of the project and then become frustrated when agency needs hold up the schedule after the fact, or after the regulatory need has been defined. They must realize that it takes time to perform all the necessary analysis and design reviews and get the necessary permits processed and that this time must be incorporated into the schedule.

HUMAN RESOURCE AND EXPERTISE OBJECTIVES

Technical, functional, and regulatory requirements are critical to any project's success but no more important than the human requirements for the project. The expertise of the personnel who will help develop, design, and install the project is likely to be varied and wide ranging. The existing expertise and experience of the project team is one requirement, but the project manager may also need to consider the availability of outside expertise. He or she may also need to consider on-site training requirements necessary to bring a fully functioning project team up to speed on site. Once the necessary expertise categories are defined, they must be quantified—how many crews or individuals are needed for the project? For example, if three welding crews are needed and only one is available at a given time, the schedule will be affected, and the effect may not necessarily be linear, because other types of crews will also be held up waiting for the welding to be completed.

The project manager must consider the types of engineers (e.g., industrial, chemical, mechanical, electrical) and analysts that are needed for their expertise during the design phase of the project's life cycle. He or she needs to obtain the input of regulatory experts; industrial hygienists; operations and maintenance experts; instrumentation experts; and human factors experts, or ergonomists, among others. One expert in each discipline is usually considered adequate; however, this is dependent upon the size and complexity of the project. It is critical to avoid delays in project implementation due to the lack of input from any of the team's members. Regular project meetings must include the presence and input of each discipline on the team, and it is the project manager's responsibility to ensure that all team members attend project meetings. It is not the objective of this chapter to provide a formula for determining the correct number of personnel for each discipline; the objective is to provide information to consider in determining the makeup of the project team (Shermerhorn 1993).

STATEMENTS OF WORK, TASK STATEMENTS, AND WORK BREAKDOWN STRUCTURE

When objectives, goals, and project requirements are defined and agreed upon, descriptions of the work to be done can then be developed. This is started through the development of a statement of work, task statements, and a work-breakdown structure. The *statement of work*, although often used interchangeably with a task statement, could be viewed more as a description of the overall work to be completed. It can also contain a complete list of task statements (or individual tasks within the overall project). Examples of statements of work are:

1. Excavation and earthwork
2. Driving piles
3. Laying concrete foundations
4. Running electricity to the site
5. Running water and steam to the site
6. Installing structural steel
7. Setting the vessels
8. Installing the pumps, exchangers, valves, etc.
9. Installing the piping, valves, etc.
10. Installing control-system wiring and computers

Each of these statements of work will include a description of any work that requires the input and coordination of the safety engineer/project manager.

Task statements are more defined, specific task descriptions within each statement of work, and they lend themselves to integration into a work breakdown structure (which will be discussed in upcoming paragraphs). Task statements associated with the excavation work of a project might be: perform a soil analysis, complete a nearby exposure and structures analysis, perform a water-content analysis, perform water influx and drainage analysis, complete a shoring or sidewall angling design, implement a spoil pile management plan, and design a confined-space-entry plan. Task statements for a "Complete Pile Driving" statement of work may require the safety engineer to evaluate the vibration impact on nearby structures and perform an evaluation of noise exposure for crews working in the area. Setting vessels, installing piping, and bringing utilities to the site may require the safety engineering project manager to evaluate layout and placement as well as accessibility and adherence to material specification and welding requirements. These are a few examples of possible task statements that may be developed in a project that has a particularly significant amount of safety engineering issues associated with it. Each task statement list is specific to the type of project being undertaken (Eisner 2002).

A *work breakdown structure* is a formal categorization of the work to be performed as part of a project. Task statements are developed under each work categorization. Organizing the project work into categories can help in establishing the project schedule and in allocating human resources—in terms of both crafts and expertise and pure worker numbers (Eisner 2002). In a safety engineering project, a work breakdown structure and associated task statements may look like the following example, which could describe the elevated-work aspects associated with the "installing the piping and valves" statement of work noted in the list above.

Work Breakdown Structure Example

1. Pre-project analyses
   1.1. Support structure analysis of integrity of members intended for use in fall-protection systems
   1.2. Determination of attachment points and weight limitations
   1.3. Determination of scaffold vs. harness and lanyard
   1.4. Support structure analysis of scaffold base if scaffolding is required
   1.5. Analysis of fall-protection equipment status and condition
2. Fall-protection system design (based on pre-project analyses)
   2.1. Scaffold design (and approval if large enough to require design oversight by a professional engineer)
   2.2. Lanyard and harness system design (allowing for adequate mobility and fall-distance limitation)
   2.3. Access system design
   2.4. Specification development for harnesses, lanyards and connections
3. Fall-protection system construction and installation
   3.1. Oversight of contractors or employee construction crews
   3.2. Inspection and testing of system upon installation completion
   3.3. Approval of necessary scaffolding permits
4. Site inspections during elevated work to determine continued compliance and condition integrity
5. Updating of fall-protection permits as needed throughout project

Most safety engineers would not be assigned as a project manager for a construction project. However, a safety engineer might be assigned to manage the safety-related aspects of an overall construction project, and all the same principles apply to the tasks and organization of the safety-engineering-related work (such as for the fall-protection example noted above). It is also important for a safety engineer to understand how a construction project is described, organized, arranged, and managed so that he or she will be able to participate in, integrate with, and contribute to its development and implementation (Eisner 2002).

SCHEDULING
PERT Charting

PERT is an acronym for *Program Evaluation and Review Technique*. This technique is often used to create the project network layout and is also referred to as a network diagram. PERT is described as a critical path analysis tool since it can help an engineer to determine which pathway through a network is critical to the project's on-schedule completion.

A safety engineer can use a PERT chart as a planning tool to lay out the steps of a project and determine a sequential relationship between activities. It allows the project safety engineer to find the optimum route through the project to achieve its objective. It can be used in assessing the risk of not completing a project on time (or to the project team's satisfaction) through a time-estimating process that assigns three levels of estimated implementation time to each activity—an optimistic time, a pessimistic time, and a most likely time. From these estimates, it is possible for the safety engineer to develop probability distributions around the completion time for each activity. From that, a risk value can be assigned to the completion of the project. From a practical standpoint, most practicing engineers use only two time-estimate levels (earliest and latest) for completion and often do not carry out probability distribution determinations. This type of time estimating is most often used to determine a project's critical path or to identify possible bottlenecks.

On a PERT chart, activity milestones are represented by *nodes* (see Figure 1) that are generally shown as positions in time using either the beginning or the end of an operation. The nodes are connected by lines called *arcs*. Each arc represents the time needed to complete the activity within the project's scope. Activities that are needed to ensure a correct sequence in a project but have zero time or no cost are referred to as *dummy activities* or *dummy nodes* (Freivalds and Niebel 2008).

A PERT network is devised and analyzed using its principles and methods as follows (as shown in Figure 1 and Table 1):

1. Determine and define known end events and milestones; ask what needs to be accomplished to achieve each objective (or milestone).
2. Work backward from each milestone (node) until reaching the project start. This will help to define the steps in the project.
3. While determining the steps in the project, this backward approach will help to define the sequential relationship of each node to each other (i.e., node A must be done before node B, node C can be done concurrently with node D). This sequence can be serial and/or parallel. Each node should be labeled—in this case, they are given letters.
4. These first three steps allow the analyst to draw a network, as shown in Figure 1. At this point he or she must estimate the amount of time, given available resources, that it will take to achieve each milestone and assign that number of hours, days, or weeks to the arc between the

TABLE 1

PERT Chart Analysis Data

Activity	Description	Task Duration	Early Start (Day)	Late Start (Day)	Slack Time
AB	Analyze excavation incidents	8 days	0	10 – 8 = 2	2 days
AD	Analyze lockout/tagout incidents	11 days	0	11 – 11 = 0	0 days
AF	Analyze confined-space incidents	3 days	0	13 – 3 = 10	10 days
BC	Create soil analysis survey and map	7 days	8	17 – 7 = 10	2 days
BD	Include excavation findings in LO	1 day	8	11 – 1 = 10	2 days
BE	Develop excavation procedures	7 days	8	18 – 7 = 11	3 days
DC	Identify and diagram energy sources	6 days	11	17 – 6 = 11	0 days
DE	Develop LO/TO procedures	5 days	11	18 – 5 = 13	2 days
DF	Dummy node	0 days	11	--	--
FE	Develop confined-space-entry procedure	5 days	11	18 – 5 = 13	2 days
FG	Develop confined-space-entry-training program	7 days	11	27 – 7 = 20	9 days
CG	Develop excavation training program	10 days	17	27 – 10 = 17	0 days
EG	Develop LO/TO training program	9 days	16	27 – 9 = 18	2 days
GH	Present complete training program to workers	5 days	27	32 – 5 = 27	0 days

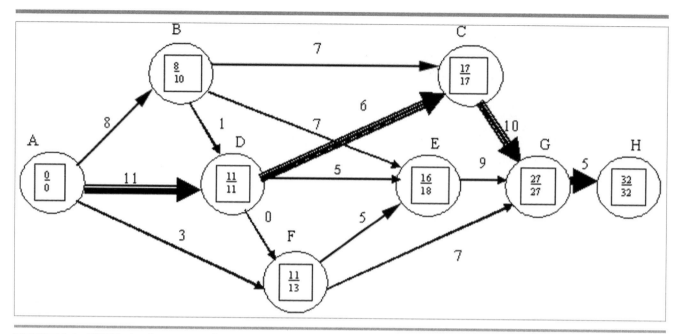

FIGURE 1. PERT chart representing a project to develop a training program based on injury-type history (critical path is highlighted with bolder arrows)

starting milestone node and its related end node. For example, in the Figure 1 network, start at node A (day 0). Achieving node B (arc AB—Analyze excavation accidents, Table 1)—is expected to take 8 days with available resources.

5. Once all the time estimates are assigned, begin at node A, and using the 8-day estimate for arc AB, the 11-day estimate for arc AD, and the 3-day estimate to achieve node F, determine the earliest possible time to achieve each milestone. This will define the earliest possible time in which each of the following nodes can be achieved. Input that number (day) inside the node above the line. Work your way through the network, being sure to respect the sequential relationship you built for the project. [Suppose you have only one welding crew and they have two steps in the project (arc-node combinations). One crew can work on only one step at time, so one step must be completed before the next can be started.]

6. When you reach the node indicating project completion, you will have defined the earliest completion time (the smallest number of days to complete the project). In the Figure 1 example, it will take 32 days to complete the project.

7. From the final node, work backward through the network, again using the completion-time estimates to determine the latest possible completion time for each node. This number is recorded below the line inside the circle of each node. In the example, the project completion (node H) is at day 32 and with only one node immediately preceding it (node G), the analyst must address only the 5 days needed to accomplish task GH and record 32 − 5 = day 27 below the line at node G. From node G, the analyst travels backward, subtracting the 10 days it will take to complete task CG from day 27 and determines that the latest completion of node C is achieved at day 17. This number is recorded below the line inside the node C circle. At this point, care must be exercised when a node has more than one arc coming from it. The analyst must record the value defined by the arc yielding the lowest value. For example, from node G, the analyst travels backward, subtracting the 7 days needed to complete task FG from day 27 and records the latest completion of node F. This is 20; however, the lowest value is achieved for node F when subtracting the 5 days it takes to complete

task FE from node E's latest value of 18, so the latest start for node F is at day 13. This number is recorded below the line inside the circle of node F. The analyst works back through the entire network in this manner until reaching day zero at the project's first task.

8. Once the earliest and latest start times are determined, the analyst can set up a table, such as Table 1, and determine slack time for each task where it applies, and most importantly, can determine the project's critical path. This is the path through the network along which the project can afford no slack time; a delay in the tasks along it will result in a delay in the project.

9. Complete the first three columns of the table (Table 1) from the information already developed as shown on the network. The earliest start day for each node is the number above the line inside the node circle—enter that in the fourth column of the table.

10. For the fifth column, labeled "Late start (day)," some additional considerations must be integrated into the entries. First, use the day value below the line in the final node and subtract the task duration leading to it. Then subtract the earliest start value for that task (the number below the line inside the project's final milestone or node). This result is the slack time or the amount of delay time the project can stand without suffering a delay in the overall project. For example, if slack time is established at 2 days and a task requires an order of materials to be received on day 22, the project will not be delayed as long as the material arrives and the task can be completed by day 24.

11. As the analyst looks down the last column, it is easy to determine the critical path from the arcs or tasks that have zero (0) slack time. In this example, the critical path shows that the tasks AD (Analyze the lockout/tagout incidents), DC (Identify and diagram energy sources), CG (Develop excavation training program) and GH (Present complete training program to workers) cannot experience any delays or the whole project will take longer than the 32 days expected.

For projects of critical time or financial demand, the most resources or most attention should be focused on the critical path tasks to ensure that they are not delayed. In giving information to the operation's decision makers, the analyst can provide these same data at three levels: optimistic estimate, expected estimate, and pessimistic estimate. The analyst can also draw confidence bands around each duration estimate. The PERT chart provides much information for decision makers to use in determining time estimates, human resource allocation, costs, and so on for achieving a project's objective (Eisner 2002; Freivalds and Niebel 2008). There are many project management software tools available on the market that provide these scheduling tools, such as *Microsoft Project*, *Tenrox Project Management*, *Matchware Mindview*, *Genius Project* for Domino, *Seavus Project Planner*, *Method 123 Project Plan*, and AEC Software's *Fast Track Schedule*. A comparison can be found at www.project-management-software-review.toptenreviews.com.

Gantt Charting

A Gantt chart is a valuable tool that is used extensively throughout industry today even though it was developed in the 1940s. It is often used in safety engineering applications as well as in many other types of project-based applications. It is a project planning and control technique that is designed to show the expected completion times for each step in a project. It makes use of a horizontal timeline. Bars are plotted against this timeline to represent first the expected and then the actual completion times for each project activity. The benefits of its use are many. First, it forces a safety engineer to plan and lay out a project before starting the work. Second, the engineer can tell from the layout whether there is activity overlap, and where this is the case, determine whether resources are available to support the overlap. Third, from this graphical representation of project status, the engineer can easily tell whether the project is ahead of or behind schedule.

Gantt charts have some disadvantages in that they don't always give the project engineer the ability to see the interactions between two or more activities—if the same group of workers for a specific craft is scheduled to do three things at once, the Gantt chart alone would not show this. For example, suppose that the same group of pipe fitters is expected to install energy-isolating blinds or blanks in three separate toxic liquid lines in three separate locations in the plant during a maintenance turnaround all at the same time. The Gantt chart would not show this. It does not point out specific manpower allocation problems, but it does at least help to provide the impetus for a project engineer to think about the problem when an overlap occurs. See Figure 2 for an example of a Gantt chart.

Bars are used to show the expected time to complete each activity in terms of start and finish dates as well as expected duration. The bars are also darkened to indicate activity completion status. In the Figure 2 project, the darkened bars indicate completion; the bar colored on the left and uncolored on the right indicates an activity that is in progress. One can see that everything is completed through September. The activities corresponding to the uncolored bars have not yet been started. From this Gantt chart, one can tell that the first five activities were completed as scheduled, as were the seventh and eighth. Completing the PHAs is in progress and still on schedule, but training, writing, and implementing the contractor safety program are not started yet. If the current date is 30 September, all is on schedule, but if it is 20 November, these four activities would be behind schedule.

One powerful benefit of a Gantt chart is that it allows a project safety engineer to determine whether there are enough resources (humans, computers, etc.) to work on overlapping activities. This is illustrated in the following example:

The safety engineer may ask if there are enough hourly employee time resources available in February to "implement employee participation" at the same time as they provide input to "writing the process safety information (PSI) document." If so, the project schedule is acceptable as is, but if a maximum-production test happens to be scheduled, which demands people time, and enough people are no longer available at the same time, the schedule will have to be adjusted. The PSI activity may have to be moved to March, but then that will impact the rest of the schedule.

This short example illustrates one of the benefits of Gantt-charting a project, but there are many. This

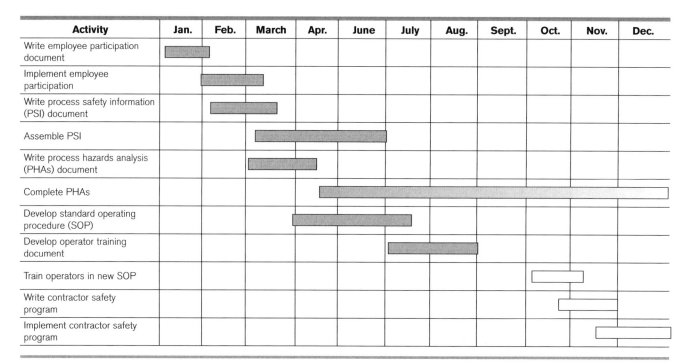

FIGURE 2. Gantt chart for developing a written process safety management program (not a complete project) Figure key: darkened color completed, fading color in progress, white is not started.

TABLE 2

Task Responsibility Matrix: Vessel and Piping Installation Tasks

	Welders		Pipe Fitters		Inspectors		Instrument Engineers		
	No. people in crew	Person-weeks	No. people in crew	Person-weeks	No. people in crew	Person-weeks	No. people in crew	Person-weeks	Total Person-Weeks
Task A	4	4	5	5	2	2	2	2	13
Task B							1	1	1
Task C	4	4	5	5	2	2			11
Task D	4	2	7	3.5	2	1			6.5
Task E	4	8	7	14	2	4	2	2	28
Task F							2	1	1
Task G							2	1	1
Task H	4	4	7	7	2	2	2	2	15
Totals		22		34.5		11		9	76.5

(Adapted from Eisner 2002)

type of schedule/chart is regularly used as a reference whenever project schedule and resource estimates are discussed.

FINANCIAL CONSIDERATIONS AND COST MONITORING

Project Budget

Once the project manager has a solid project plan in place and the work is defined, he or she must determine a budget for the project and a means to determine the financial performance and budget status throughout the life of the project. This process includes developing a task responsibility matrix, developing a direct labor and materials budget from the matrix, and then determining a cost budget per week. An example is the best way to illustrate the process. The sour gas processing plant project described above will be used to show the budget development steps. The example will not provide specific costs for each budget element, as this responsibility will reside with individual project managers and their staff analysts. The cost (in dollar amounts) used in this example are for illustration purposes only.

From Table 2 we determine that the project will require 22 person-weeks of work for the welding crew, 34.5 for the pipe fitters, 11 for the inspectors, and 9 for the instrument engineers. The total human-resource demand for the project is 76.5 person-weeks. The chart shows the person-weeks by task as well.

The process of establishing a project budget and analyzing the project's financial performance on a weekly basis (or other appropriate frequency), allows the project manager to maintain fiscal control early enough in the project so that available funds do not run out before the project is complete. There are several means for assessing budget performance; however, this tabular method is a relatively straightforward means that does not require extensive financial training (see Tables 3 and 4). It should be noted that the examples used in this chapter propose cost percentages

TABLE 3

Project Budget: Vessel and Piping Installation Tasks

Direct Labor	Rate/Week (in $)	Person-Weeks Required	Cost (in $)
Welders	1400	22	30,800
Pipe fitters	1120	34.5	38,640
Inspectors	1200	11	13,200
Instrumentation engineers	1500	9	13,500
Subtotal 1			96,140
Fringe rate @ 30%			28,842
Subtotal 2			124,982
Overhead rate @ 68%			84,988
Direct costs (materials, supplies, delivery, etc.)	vessels, piping, welding supplies		526,700
Subtotal 3			736,670
General and administrative @ 12%			88,400
Total cost			825,070

(Adapted from Eisner 2002)

TABLE 4

	Cost Budget by Week (in thousands of dollars)													
	Week 1		Week 2		Week 3		Week 4		Week 5		Week 6		Week 7	
Cost category	Person-weeks	Cost ($)	Person-weeks	Cost ($)	Person-weeks	Cost ($)	Person-weeks	Cost ($)	Person-weeks	Cost ($)	Person-weeks	Cost ($)	Person-weeks	Cost ($)
Welders	4	5.6	--	--	4	5.6	2	2.8	8	11.2	--	--	4	5.6
Pipe fitters	5	5.6	--	--	5	5.6	3.5	3.92	14	15.7	--	--	7	7.84
Inspectors	2	2.4	--	--	2	2.4	1	1.2	4	4.8	--	--	2	2.4
Instrumentation engineers	2	3.0	1	1.5	--	--	2	3.0	1	1.5	1	1.5	2	3.0
Fringe @ 30%	--	4.98	--	0.45	--	4.08	--	3.3	--	9.95	--	0.45	--	5.65
Subtotal 1	--	21.58	--	1.95	--	17.68	--	14.2	--	43.15	--	1.95	--	24.5
Overhead @ 68%	--	14.7	--	1.33	--	12.0	--	9.7	--	29.4	--	1.33	--	16.65
Subtotal 2	--	36.3	--	3.28	--	29.7	--	23.9	--	72.6	--	3.28	--	41.15
Materials and supplies	--	95.14	--	10	--	95.14	--	105	--	96	--	10	--	115.42
Subtotal 3	--	131.5	--	13.3	--	124.8	--	128.9	--	168.6	--	13.3	--	156.6
General and administrative @ 12%	--	15.8	--	1.6	--	14.9	--	15.5	--	20.2	--	1.6	--	18.8
Total cost	--	147.3	--	14.9	--	139.7	--	144.4	--	188.8	--	14.9	--	175.4
Cumulative cost	--	147.3	--	162.2	--	301.8	--	446.2	--	635.1	--	650.0	--	825.2

(Adapted from Eisner 2002)

that are only for example purposes. Anyone using these types of tables should use only their company's internal cost data.

Engineering Economics

Engineering economics is an area used by most engineers and project managers to determine specific project options and to monitor the progress of the project performance. For more information on this subject, as well as general economic analysis, refer to the opening chapter in this handbook as well as several of the cost analysis and budgeting chapters in each topic area. Engineering economics will be treated here briefly in the context of managing safety engineering projects. Some of the concepts considered include:

- engineering economic decisions
- equivalence and interest formulas
- loan transactions
- annual equivalent worth analysis
- internal rate of return
- depreciation
- minimum acceptable rates of return
- lease versus buy decisions
- inflation analysis
- capital budgeting
- bond investments
- interest rates
- present and future worth analysis
- rate of return analysis
- return on investment (ROI)
- taxes
- developing project cash flows
- replacement decisions
- project risk and uncertainty
- economic equivalence

These tools allow project managers to make informed decisions about projects from a financial, risk, and uncertainty point of view. While each of these concepts, principles, and analytical techniques is used in managing a project, this chapter will address only a few of them to avoid overlap with other chapters in this handbook.

The Time Value of Money

The *time value of money* is an important concept in engineering economics and bears discussion and a review of the equations that are used to determine it.

Some of the following equations will be used in a subsequent example:

Single Flow

Convert present value to future value:
$$F = P(1 + i)^N \quad (1)$$

Convert future value to present value:
$$P = F(1 + i)^{-N} \quad (2)$$

Equal Payments

Convert periodic annuities (monthly or annual payments) for a given period to a future value:
$$F = A[(1 + i)^N - 1/i] \quad (3)$$

Convert a future value to annuities (monthly or annual values) for a given period:
$$A = F[i/(1 + i)^N - 1] \quad (4)$$

Convert periodic annuities (monthly or annual payments) for a given period to a present value:
$$P = A[(1+ i)^N - 1/i(1 + i)^N] \quad (5)$$

Convert a present value to annuities (monthly or annual values) for a given period:
$$A = P[i(1 + i)^N/(1+ i)^N - 1] \quad (6)$$

where:
- P = present value
- F = future value
- A = annuity
- i = interest rate
- N = number of compounding periods

The example will be the determination of the cash flow for each of two projects in which a decision will be made either to develop a new project or to invest in the upgrade or improvement of an existing facility. A comparison will be made by converting cash flows for each project to a *net present value* (also referred to as *net present worth*) (Park 1997).

The decision to upgrade an existing methanol (MeOH) storage facility (Defender) or to design and install a new methanol storage facility (Challenger) is both a safety and an economic decision. One method for making this decision is shown in the "Question for Study" sidebar.

QUESTION FOR STUDY

Applying Engineering Economics

Some crude oils are difficult to keep flowing in temperatures below zero. Hydrate salts can precipitate out of the liquid phase as the oil gets further from the well and cools. These salts create a dangerous condition as they form plugs in the line. This traps high pressure and creates a condition that results in a projectile being rocketed down the pipeline as one of the hydrate plugs dislodges and releases the trapped pressure behind it. There is a risk of pipeline damage, potential physical injury, exposure to hydrocarbon, and environmental damage.

A method for preventing this trapped pressure condition is to inject methanol (MeOH) into the oil stream. This keeps the hydrate salts from precipitating out and thus keeps the line from plugging. The present methanol storage and loading facility is manually controlled, with no fire protection and deteriorating tanks. Methanol leaks occur frequently. The scope of repairs and upgrades to this system is extensive. The storage tanks are rusting and leaking at the riveted joints. The manual-level control system allows frequent overfills. There is no fire water available at this site and MeOH is flammable. The present system has been in service for 5 years. Due to permit requirements, upgrades are required to achieve minimum acceptable environmental and safety standards. Upgrades costing $134,000 will extend the system's life for five more years. However, upgrades will not completely stop the leaks. The expected leak losses will amount to $5000/year and the clean-up costs will be another $5000/year. The annual operating costs for this upgraded facility are expected to be $36,000 and maintenance costs are expected to be $24,000/year. Revenues generated would be realized from operational efficiency improvements. Savings are due to reduced risk and reduced compliance costs. The amounts would be:

Year 1 = $158,000
Year 2 = $160,000
Year 3 = $140,000
Year 4 = $137,000
Year 5 = $126,000.

A project design team has proposed an entirely new facility design with input from a safety analysis team. The design meets all acceptable environmental and industry standards and practices, including appropriate fire protection and level control system. The new facility, which requires an investment of $325,000, would last 5 years before a major upgrade would be required. However, it is believed that in five years oil transfer technology will be developed to the point that methanol will become obsolete (pipeline heating systems and insulation material). Management does not expect any spill clean-up costs; however, they would like a $5000 per year contingency set aside for education of the community (on MeOH safety), possible clean up, possible evacuation, etc., just in case. The annual operating and maintenance costs would be $12,000 and $6000 respectively. For this new system, revenues generated would be realized from operational efficiency improvements. The savings are due to reduced risk and reduced compliance costs. The amounts would be:

Year 1 = $180,000
Year 2 = $170,000
Year 3 = $160,000
Year 4 = $150,000
Year 5 = $140,000.

Even though a methanol system is expected to become obsolete, some of the parts in this system could be salvaged at the end of five years. The equipment should be worth $10,000. If this company's minimum acceptable rate of return is 15 percent, its tax rate is 40 percent, and the depreciation class for each system is 5-year class Modified Accelerated Cost Recovery System (MACRS), should this company upgrade the old facility or build a new one?

Net Present-Worth Analysis—Cash Flow Statement—Defender (Existing System)

			5-Year MACRS Depreciation			
	0	1	2	3	4	5
Income Statement		(20%)	(32%)	(19.2%)	(11.52%)	(11.52%)
Revenue						
Savings, reduced risk		$158,000	$160,000	$140,000	$137,000	$126,000
Expenses						
• Operating costs		$36,000	$36,000	$36,000	$36,000	$36,000
• Maintenance costs		$24,000	$24,000	$24,000	$24,000	$24,000
• Materials (losses)		$5000	$5000	$5000	$5000	$5000
• Spill clean-up		$5000	$5000	$5000	$5000	$5000
• Depreciation		$26,800	$42,880	$25,728	$15,436	$15,436
Taxable Income		$61,200	$47,120	$44,272	$51,564	$40,564
• Income tax (40%)		$24,480	$18,848	$17,709	$20,625	$16,225
• Net income		$36,720	$28,272	$26,563	$30,938	$24,339
Cash Flow Statement	0	1	2	3	4	5
Operating Activities						
• Net income (A)		$36,720	$28,272	$26,563	$30,938	$24,339
• Depreciation (B)		$26,800	$42,880	$25,728	$15,436	$15,436
Investment Activity						
• Investment (I)	($134,000)					
• Salvage (S)						$0
Net Cash Flow (A + B + S)		$63,520	$71,152	$52,292	$46,374	$39,775

$$NPW = I + F_1(1 + i)^{-N} + F_2(1 + i)^{-N} + F_3(1 + i)^{-N} + F_4(1 + i)^{-N} + F_5(1 + i)^{-N}$$

$$NPW = \$134,000 + \$63,520(1 + 0.15)^1 + \$71,152(1 + 0.15)^2 + \$52,292(1 + 0.15)^3 + \$46,374(1 + 0.15)^4 + \$39,775(1 + 0.15)^5$$

$$NPW = -\$134,000 + \$63,520(0.8696) + \$71,152(0.7514) + \$52,292(0.6575) + \$46,374(0.5717) + \$39,775(0.4971)$$

$$NPW = -\$134,000 + \$55,236 + \$53,463 + \$34,381 + \$26,512 + \$19,772$$

NPW = $55,364

Managing a Safety Engineering Project

Net Present-Worth Analysis—Cash Flow Statement—Challenger (New System)						
		5-Year MACRS Depreciation				
Income Statement	0	1 (20%)	2 (32%)	3 (19.2%)	4 (11.52%)	5 (11.52%)
Revenue						
Savings, reduced risk		$180,000	$170,000	$160,000	$150,000	$140,000
Expenses						
• Operating costs		$12,000	$12,000	$12,000	$12,000	$12,000
• Maintenance costs		$24,000	$24,000	$24,000	$24,000	$24,000
• Materials (losses)		$0	$0	$0	$0	$0
• Spill clean-up (contingency)		$5000	$5000	$5000	$5000	$5000
• Depreciation		$65,000	$104,000	$62,400	$37,440	$37,440
Taxable Income		$92,000	$43,000	$74,600	$89,560	$79,560
• Income tax (40%)		$36,800	$17,200	$29,840	$35,824	$31,824
• Net income		$55,200	$25,800	$44,760	$53,736	$47,736
Cash Flow Statement	0	1	2	3	4	5
Operating Activities						
• Net income (A)		$55,200	$25,800	$44,760	$53,736	$47,736
• Depreciation (B)		$65,000	$104,000	$62,400	$37,440	$37,440
Investment Activity						
• Investment (I)	($325,000)					
• Salvage (S)						$0
Net Cash Flow (A + B + S)		$120,200	$129,800	$107,160	$91,296	$95,176

$$NPW = I + F_1(1 + i)^{-N} + F_2(1 + i)^{-N} + F_3(1 + i)^{-N} + F_4(1 + i)^{-N} + F_5(1 + i)^{-N}$$

$$NPW = -\$325,000 + \$120,200(1 + 0.15)^1 + \$129,800(1 + 0.15)^2 + \$107,160(1 + 0.15)^3 + \$91,296(1 + 0.15)^4 + \$95,176(1 + 0.15)^5$$

$$NPW = -\$325,000 + \$120,200(0.8696) + \$129,800(0.7514) + \$107,160(0.6575) + \$91,296(0.5717) + \$95,176(0.4971)$$

$$NPW = -\$325,000 + \$104,525 + \$98,146 + \$70,457 + \$52,193 + \$47,311$$

$$NPW = \$47,636$$

Final Result and Decision:

The company should choose the Defender. Its net present worth is $55,364 versus the Challenger's net present worth of $47,636.

LEADERSHIP AND PROJECT MANAGEMENT

Attributes of Leaders and Project Managers

It has been said that people usually do not get fired for a lack of technical skills; they lose their jobs more often because of their lack of interpersonal skills. A person can be a project manager by title but may not be an effective leader. An effective leader without sound technical skills may still succeed; but to do so, he or she must be strong on interpersonal skills. A manager may be strong at handling the details of day-to-day project management but unable to achieve success due to a lack of execution when team support is missing. Schermerhorn (1993) highlights some important points from a speech by management consultant Abraham Zaleznick to a group of business executives about the differences between leaders and managers. Zaleznick said, "Leaders . . . can be dramatic and unpredictable in style. They are often obsessed by their ideas, which appear visionary and consequently excite, stimulate, and drive other people to work hard and create reality out of fantasy. They often create an atmosphere of change." About managers, Zaleznick said, "Managers . . . are usually hardworking, analytical, and fair. They often have a strong sense of belonging to the organization and take pride in maintaining and improving the status quo. They tend to focus on the process, while leaders focus on substance." Leaders are not managers, and managers are not necessarily leaders, but for a project to succeed they should be (Schermerhorn 1993).

Project managers are expected to accomplish project objectives through effective planning, organizing, directing, and monitoring. The planning and organizing can be done in the office with the door closed, but when it becomes necessary to move on to directing and monitoring, interacting with people is inevitable (Eisner 2002). Schermerhorn (1993) explains, "Great leaders . . . get extraordinary things done in organizations by inspiring and motivating others toward a common purpose." In order for project managers who are good leaders to succeed in the directing and monitoring, they must be effective motivators, communicators, and team players, as well as strong in interpersonal relations. They must also be strong in managing the dynamics of a group with all of its interactions, friction, conflict, synergism, and antagonism.

Eisner (2002) suggests twenty attributes that a project manager should have to succeed. While it would be difficult for any one person to have all twenty, a project manager should make a strong effort to develop these attributes:

- communicates well and shares information
- delegates appropriately
- is well-organized
- supports and motivates people
- is a good listener
- is open-minded and flexible
- gives constructive criticism
- has a positive attitude
- is technically competent
- is disciplined
- is a team builder and player
- is able to evaluate and select people
- is dedicated to accomplishing goals
- has the courage and skill to resolve conflicts
- is balanced
- is a problem solver
- takes initiative
- is creative
- is an integrator
- makes decisions

These attributes are likely to be desirable in anyone; however, a manager who has a high percentage of them is likely to be a successful project manager. Although Eisner (2002) defines each attribute further, they are not defined here so that readers can consider the attributes from their own perspective and consider how and how much they might improve their own project management performance.

Leadership Self-Evaluation

While it is left up to readers to assess themselves as to adherence to behaviors supportive of these attributes, Eisner (2002) proposes a method to assess one's own leadership and managerial scores. He suggests using the twenty attributes listed in the preceding section and ranking oneself for each attribute using a scale of 0 to 5, with 5 being "almost always," 4 being "most of the time," 3 being "often," 2 being "sometimes," 1 being "rarely" and 0 being "never." With a score of 80 to 100, you are likely to be a good project manager. With a score between 60 and 79, you are doing well but should seek ways to improve in the areas where you scored lower. If you scored in the 40-to-59 range, you may still be considered for a project manager position, but you may need more training and/or experience. A score below 40 indicates that you may need extensive work before becoming a project manager.

Eisner (2002) also provides the attributes of a leader, which come from his survey results. They are broken down into *critical*, *extremely important*, and *significant* attributes:

Critical

1. Empowering, supporting, motivating, trusting
2. Having a vision—a long-term viewpoint
3. Cooperating, sharing, team playing, and team building
4. Renewing, learning, growing, educating

Extremely Important

1. Being communicative
2. Having culture and values, serving as a role model
3. Being productive, efficient, determined

Significant

1. Demonstrating time management
2. Being action-oriented
3. Making a contribution, commitment, legacy
4. Being innovative, imaginative
5. Having integrity, morality, humanity
6. Demonstrating skill, knowledge, substance

One may want to consider using the same scoring mechanism for these leadership attributes as suggested for the managerial attributes. With thirteen attributes, the maximum score is 65; someone with aspirations of becoming a good leader and good manager should seek to score in the 50-to-65 range.

Situational Leadership

Hersey and Blanchard (1977) introduced the concept of *situational leadership*—leaders changing and adjusting their style to fit the situation and the people involved. Situational leadership is a well-established model built around the concept that each situation requires the application of a combination of two possible behavior dimensions—task or directive behavior and relationship or supportive behavior (Hersey and Blanchard 1977). In each case, a leader has to determine which combination is required by the situation and then correctly apply the appropriate behavior to properly manage the situation. They use four situations to describe the fundamentals of their model:

Situation 1 (S1): High task, low relationship—leader assumes *telling* role

Situation 2 (S2): High task, high relationship—leader assumes *selling* role

Situation 3 (S3): High relationship, low task—leader assumes *participating* role

Situation 4 (S4): Low relationship, low task—leader assumes *delegating* role

In an S1 situation, subordinates are usually new to a task and do not know how to do it. At this stage they don't know what they don't know. They need to be *told* what the task is and how to do it. They don't necessarily need a close relationship with the leader. In an S2 situation, workers are thought to be developing some competence, and now at least they know what they don't know. Because of this, they begin to develop more of an interest but still have to rely on the leader for guidance, so the need for a closer relationship is there. Developing workers want to know what the leader knows. In an S3 situation, workers have developed confidence and competence. They can handle the situation or task without input (task direction is not necessary); however, it is, figuratively speaking, the first time on their own, so they would like input and feedback about their performance and need a close (or *high*) relationship with the leader. In S4 situations, workers are fully developed and know how to handle the situation or perform the task without input and do not need feedback. Task-direction need and relationship with the leader are both low (Hersey and Blanchard 1977).

A leader must be able to constantly assess this very dynamic process and correctly determine in which of these four categories workers or situations are. Leadership behavior must be applied accordingly. In an S4 situation, a leader can delegate a task and stand back without involvement, but if a leader gives S4 workers detailed instructions on how to do a task, they will become frustrated. If the leader delegates a task to S1-level people and walks away without instruction or direction, there will also be frustration and task performance problems. Correct application of the appropriate leadership behavior is critical and constantly changing. People who subscribe to this leadership model cannot stay with one leadership style and apply it in all cases. If they do, team or project performance will likely be low (Hersey and Blanchard 1977).

Team Building and Interaction

One who manages a project knows the value of a contributing project team that works effectively together. Before teamwork is discussed, it will be of value to define some of the basic aspects of teams. A team is generally considered to be a group of people (more than two) gathered together to accomplish an established objective through regular interaction and input. *Interaction* is the critical word in this definition, as this is where we apply the word *teamwork*. Teamwork might be defined as the process of coming together in this interaction. It is this interaction that can produce significant accomplishment, but it can also be the greatest contribution to project failure if the interaction is not managed properly. It is generally recognized that a team, through synergy of ideas and talents, can produce products of much greater value than if each individual on the team worked alone toward the same goal. Unfortunately, many things get in the way of this positive synergy, and the project manager is challenged to keep it functioning (Eisner 2002, Nahmias 1993).

There are many safety engineering examples in which multidisciplinary teams working together are absolutely essential to preventing incidents. Any type of complex hazard analysis requires the input of many disciplines and many types of expertise. When a hot work project or a confined-space-entry or excavation project is proposed, a multidisciplinary team is necessary to plan the work as well as to implement it. Before we look at a specific safety engineering project, we will illustrate teamwork with a sports example. Lack of teamwork on the basketball court can create problems; the most evident problem is losing the game. Suppose a playground basketball star playing on a formal basketball team hogs the ball, takes all the shots without passing to teammates, and does not try to get others in a position to take shots or get the ball. He cannot beat a team of five on the basketball court, but because of selfish play, he becomes alienated from his other four team members. They then either take steps to isolate the selfish player or stop contributing. Both courses of action are a detriment to the performance of the team. A coach (leader) must recognize selfish play and do what is necessary to get everyone playing together, or the team will lose games.

In a safety engineering example, a process hazards analysis team (such as a hazard and operability study team—HAZOP) must work together, because the expertise of a team of many disciplines is necessary to thoroughly study a complex system to identify its hazards. As discussion takes place during the study, ideas develop through discussion among team members, each relying on the expertise and experience of other members. Team members confirm their opinions with teammates, the final conclusions are usually well thought out, and the thoroughness and accuracy of the study is high. If one team member thinks he or she knows everything and monopolizes the conversation, other team members either discount what that person says or they themselves keep quiet and do not contribute. Both courses of action are a detriment to the performance of the team. The study facilitator (leader) must recognize that this is happening and redirect the energies of the know-it-all toward the objectives of the team.

Teams provide great benefit to an organization. According to Schermerhorn (1993), teams can provide this benefit through:

- increasing resources for problem solving
- fostering creativity and innovation
- improving quality of decision making
- enhancing members' commitment to tasks
- raising motivation through collective action
- helping control and discipline members
- satisfying individual needs as organizations grow

Teams can be ineffective when individual differences are not embraced, when tasks are poorly designed, when team members are not prepared to work or are not committed to accomplishing the objective, or when the team process is not strong—either communication is poor, the decision-making process is not well defined or established, or there is no conflict resolution mechanism (Schermerhorn 1993). According to Eisner (2002), there also may be team members who are referred to as *team busters*. These people exhibit any of these behaviors:

- question the authority of the project manager on every issue
- challenge the technical approach of the project manager
- do not follow agreed-upon decisions
- consistently go over the boss's head
- try to monopolize meeting agendas
- attempt to embarrass or challenge the project manager in front of others
- try to create a "we and they" mentality

Teams are also subject to a phenomenon called *groupthink*. According to Schermerhorn (1993), this phenomenon can be characterized by:

- having illusions of group invulnerability—the team is above criticism
- rationalizing unpleasant and disconfirming data—refusal to accept contradicting data
- believing in inherent group morality—the team is above the reproach of others
- stereotyping competing teams as weak or stupid
- applying direct pressure to those with different points of view to conform to team wishes
- self-censoring by members—they won't accept that the team may be wrong
- having illusions of unanimity—accepting consensus prematurely
- mind guarding—protecting the group from hearing disturbing or competing ideas from those not on the team

Project managers must ensure that, given the problems of team interactions, they manage the personality differences and friction effectively to achieve and maintain a high performance. They must listen to what team members are saying, should work with and spend time with team members as much as possible and feasible, and should encourage all participants to work together and contribute. Project managers should meet with key team members regularly and frequently, integrate the input of team members into the flow of the project work, and talk with the project "customer" as well as other project support people regularly and often. As the leader, the project manager should maintain a positive attitude, be supportive of all team members, and offer to help team members communicate with each other when and where necessary (Eisner 2002).

According to Schermerhorn (1993), characteristics of a high-performing project team that project managers should seek to build and achieve through their effective leadership are:

- working toward a clear and team-elevation goal
- a task-driven, results-oriented team structure
- competent, competitive team members who are willing to work hard
- a collaborative climate where people like and expect to work together
- high standards of excellence
- external support and recognition from the rest of the organization
- strong principled and ethical leadership

Eisner (2002) suggests several activities for leaders that support many of Schermerhorn's characteristics of a high-performing project team. They include:

- Develop and maintain a personal plan for team building and for operation.
- Hold periodic as well as special team meetings.
- Clarify missions, goals, and roles.
- Run the team in a participative and, where possible, consensual manner (consensus can mean that everyone does not necessarily agree with a decision, but can live with it).
- Involve the team in situation analysis and problem solving.
- Give credit to active, positive team members and their contributions.
- Assure team efficiency and productivity.
- Obtain feedback from team members.
- Integrate, coordinate, facilitate, and assure active information flow up and down, as well as sideways.
- Maintain effective communication.

Since in all cases we are talking about humans, the information presented here is for project managers and readers to think about, refer to, try to apply, and try to live by, but not to take as a guarantee for one hundred percent maximum project team performance all of the time. Managing a project and a project team is a dynamic process, and it is affected by emotions, motivations, conflict, and the general pressures of daily life. These factors contribute to making project management a challenging and sometimes frustrating experience.

Managing Conflict

No matter how project managers try to keep a team focused and working effectively toward achieving an objective on time and within budget limitations, there will be conflict between team members that can impact the team's performance. Conflict on a team is natural and can be directed toward the positive benefit of the team. When conflict arises, the project manager must resolve it by eliminating it or directing it positively.

Conflict arises in a number of areas, some of which are impersonal, such as those involving scheduling, budgets, or procedures. These are usually easier to resolve than those involving emotional attachment to a conflict position. Conflict areas such as priorities, personalities, or technical opinions are personal, and since team members may be emotionally attached to their positions, especially when personalities and opinions are involved, these types of conflicts may be more difficult to resolve. There is no one formula leaders should use to resolve conflict on the project team in all situations. In fact, the same method used to resolve one type of conflict between two team members may make the conflict worse if the same method is used for two different people. Leaders are in the difficult position of having to recognize the type of conflict as well as how each participant responds to managerial resolution action before determining the right conflict-resolution method (Eisner 2002).

There are a number of approaches to conflict resolution, and some of them are an indication of the personality of the leader trying to help resolve the conflict. If there is a conflict between two (or more) team members, project managers can choose (or will naturally resort to) the *forcing approach*, where they use their position to order the immediate resolution of the conflict. This will probably resolve the conflict temporarily, but participants are likely to remain frustrated and avoid the issue. Parents sometimes use this method when they force their fighting children to "shake hands and make up." Sometimes the conflicting parties remain angry and can respond with greater force at a later date.

A *compromising* or *sharing* resolution style means working with participants to find a mutually agreeable position that both parties can live with. This is traditional negotiating; many times both parties must give up something, and the losses may cost the project in the long run. Care should be used with this approach to ensure that the compromise solution is also optimized for the sake of the project.

An approach used by weak project managers is the *avoidance* or *withdrawal* style. This is not really considered a conflict-resolution approach; it is considered a management style, but involves simply pretending the conflict doesn't exist. Great care should be taken by upper-level management to keep this from happening, especially when project managers are new to a leadership position.

Another resolution approach is the *accommodation* or *smoothing* approach, in which the resolution is acknowledged, but its seriousness is deemphasized. This should be considered only a temporary measure to reduce the impact of the conflict at the particular moment that it surfaces, in case there is anger or too much emotion involved to deal with it in the heat of battle. It is a perfectly legitimate means to temporarily resolve a conflict (with emphasis on *temporarily*) during the process of permanent resolution when that can be accomplished professionally and when clearer heads prevail.

Probably the most effective approach to resolving conflicts is the *collaborating* or *problem-solving* approach. Team members in conflict are brought together in an analytical and problem-solving environment and collaboration becomes necessary to resolve the issue. For this approach to work, leaders (in this case, they are often facilitators) must listen closely to both sides and encourage the participants to listen to each other's position. When each position is clearly understood by both sides, misunderstandings are often highlighted and become much easier to address. If the conflict is approached as a problem, it can be made less personal and can more easily be dealt with analytically, with less emotion, and in a collaborative way. This is the recommended approach to resolving conflict on a project (Eisner 2002).

Each conflict will bring with it its own set of complexities and personalities, and there is no one right answer or tool that will be successful in resolving conflicts every time. However, if the conflict's resolution is approached through collaboration, analysis, and conversation between team members, it is likely to be more effective and will probably result not only in resolution of that conflict, but also in better project performance.

Motivation and Incentives

Many safety engineers find themselves involved in some way with motivation and incentive programs. While it is not the objective of this chapter to argue the merits of these types of programs, it is still important to discuss the role that motivation and incentives play in managing safety engineering work. This chapter will not address motivation or incentives for the purpose of working safely; it will address them from the vantage point of motivating workers and providing incentives to ensure overall project performance. Safety performance is one component in that overall measure, but safety incentive programs will not be specifically addressed.

Motivation, according to Schermerhorn (1993), is a term that refers to forces within people that drive them toward a specific goal, and it explains a level of "direction and persistence of effort expended" that people have inside for the accomplishment of a goal (in this case, work-related objectives). The impression most people would agree with is that motivated people work harder to achieve their goals than unmotivated people (Schermerhorn 1993). *Incentives* are thought to increase the level of motivation present in people.

Abraham Maslow helped the world understand the idea of motivation with his Hierarchy of Needs theory. He introduced the concept of higher- and lower-order needs and suggested that people's lower-order needs must be satisfied before someone can try to motivate them by addressing higher-order needs. Maslow tells us that physiological needs, such as rest (adequate breaks), physical comfort on the job, and reasonable work hours must be satisfied before safety needs such as safe working conditions. Both need to be satisfied before social relationships, such as friendly coworkers, interaction with customers, and a pleasant supervisor, can be satisfied. Physiological needs are considered lower-order needs, and once they are all satisfied, we should be able to motivate people by appealing to their higher-order needs, such as esteem and self-actualization (Schermerhorn 1993).

Schermerhorn saw this phenomenon firsthand during a six-year job assignment in one of the former Soviet republics. Team meetings always started out with discussions of project status and upcoming work needs, but they would often deteriorate into discussions about the problems team members were having with their living quarters, their food, lack of heat in the buildings in the winter, and so on. After much frustration on the part of management, it became apparent that it was important to listen intently to and help solve the lower-order needs of the workforce. It was an important revelation to understand that in order to get the project team focused on the work, it was necessary to first ensure that their living quarters, food, and heat were adequate. It was a classic Maslow example in real life.

Provided the project team is adequately compensated for their normal positions and their lower-order needs have been met, some of Maslow's higher-order needs could be addressed by motivation efforts. Important issues should be considered when determining the level of motivation of the team or when determining whether more motivation is needed and, if so, how much more. Almost everyone wants interesting work, and they would like it to challenge their ability to think. If a position can keep a person in the active thinking mode for most of the day, some motivation is already built into the job. If the position causes a person to switch to a mode referred to as *habits of mind* (Louis and Sutton 1991), the chances are greater that performance will be low and additional motivation will be needed (Eisner 2002).

The project team members must be treated fairly and equitably. Many people want to be recognized as being part of a team, and they show their affiliation proudly. One only needs to take a walk through a busy airport on a heavy business travel day to see all the company logos on bags and bag tags, shirts, hats, and coats to realize that team affiliation is important to people. They are proud of their employer. In addition to treating team members fairly and equitably, it is important that the leader or project manager show appreciation for what the team is accomplishing and publicly recognize its accomplishments. When there is an opportunity, the leader should also celebrate the team's successes somehow, such as by hosting a dinner or a trip to a conference or training seminar or with physical rewards (Eisner 2002).

Incentives can serve to motivate workers, and although many incentives are associated with pay for production, the space constraints of one chapter in a large reference manual limit discussion to basic principles of incentives with detailed discussion of a few. Some key principles of incentive programs that should be considered, according to Niebel and Freivalds (1999) are that the incentive program be simple, fair, and based on proven standards. They also suggest that it provide individual incentives above base pay rates and tie the incentives directly to increased production, improved safety performance, and better product quality.

The incentive can be psychological, financial, or both, but it must be meaningful to the worker (Niebel and Freivalds 1999). Care should be taken to understand individual workers in this area, because what is considered meaningful to one worker may not be meaningful to another. This author was very strongly given this message early in his career by approximately 80 percent of the women in his workforce as he passed out the 1982 safety incentive award. The award was a very large cowboy-style belt buckle that, in hindsight, was not considered meaningful by these women. Population stereotypes are important considerations when developing incentive programs. Another important consideration with incentive programs is the fact that their effect is thought to deteriorate over time—interest in any program tends to wane after the newness wears off. Also, receipt of any incentive reward long after the *rewardable* performance may make the incentive less meaningful.

According to Niebel and Freivalds (1999), a motivational climate must be present for a sound incentive program to work. This is one that supports the notion that workers want to work, want to contribute to achieving the project's stated objectives, and want to be properly and adequately rewarded for what they contribute. This type of climate is supported when the project objectives are clearly stated, supported, and realistic and when they emphasize both measurable quantity and measurable quality of output. Workers must feel like the work is their responsibility, that their supervisor is there to support their effort, and that he or she provides frequent feedback.

Some types of incentive plans are those based on piecework, measured daywork, gain-sharing, profit-sharing, and employee stock ownership (Niebel and Freivalds 1999). For more detailed discussion on pay-related incentive programs, please refer to *Methods Standards and Work Design* by Niebel and Freivalds (see reference section). According to Lack (1996), incentive programs can be established to address and recognize outstanding safety results and performance (which can be extrapolated to include production or project performance) or to address and recognize specific significant acts of good performance.

Motivating people through incentives and the work itself is a complex process and should not be taken lightly. If a program is not developed or implemented correctly, it can be worse than no program at all. If it is properly developed, actively implemented, well supported by management, and dynamic, the results can be dramatic and positive. Haight et al. (2001b), Haight and Thomas (2003), Iyer et al. (2004), and Iyer et al. (2005) show that an awareness, incentive, and motivational program had a significant effect on the rate at which loss incidents occurred in an oil production operation and the forestry division of a power company.

Much has been written on the subjects of motivation and incentives. Anyone wishing to develop a motivational or incentive plan should read the available literature and plan the program well ahead of its implementation.

Work and Workforce Analysis

Managers are often concerned about the level of performance their employees are achieving. Many standards have been established and techniques used to determine whether performance is acceptable. One aspect of performance that managers seem to pay significant attention to is the length of time it takes to complete a task or to finish a project. It is important to know how long each task will take to complete because human time costs money, and task duration will have a bearing on the total cost of the project. For this reason, it is critical that criteria be established for each task that is part of a safety engineering project. There is discussion of performance ratings in this chapter, and one of the challenges in this area is to define what is considered *normal* performance.

Allocation of Human Resources

Almost every project involves crews of people working throughout the project area during several days and under varying conditions. Except for the smallest of projects, it would be difficult for a project manager to keep up with what everyone on the project is doing. Since in many cases labor is one of the most expensive resources associated with a project, it is important for a project manager to understand how to quantify and manage the project's human resources. It is also critical that the project manager ensure that workforce crews are efficiently allocated in sufficient numbers to the most appropriate tasks for efficient execution of the project work.

Many construction and maintenance turnaround activities are defined by a particular design or repair task. Often these are well known and frequently used, so an established procedure has already been developed and manpower requirements established. Determining the manpower allocation for these tasks, takes only the knowledge and experience of a person well versed in the type of construction or maintenance being performed. Manpower requirements for many of these tasks have been defined through several years of observation and analysis of time-study and time-accounting data. Time standards have been established with efficiency and cost in mind. Because of this, many project tasks seem to push the limits of human capacity if allowances are not built in. Time studies and allowances will be discussed later in the chapter. In any case, adherence to time and performance standards will, for cost estimating, allocating resources, and implementing tasks, likely result in lower costs.

For projects and tasks that are not so stable and well established (such as safety-related program activity), one has to rely on the concept of an input-output model to measure the process and analyze the data. This should result in understanding the relationship between the quantified task activities as the input and the performance dimension variation as the output. Once the mathematical relationship is understood, one can use it to change task activities in terms of the extent of the use of available manpower to achieve task performance outcomes. Usually, higher output is desired, but the best approach to achieve that is not necessarily to bring in more people. The choice of activities can be determined through this approach as well.

Haight et al. (2001a, 2001b) and Iyer et al. (2004) determined that manpower allocation to implement a safety and health program can be designed just as any other engineered human–task-based system can. The model shown in Figure 3 depicts a safety and health

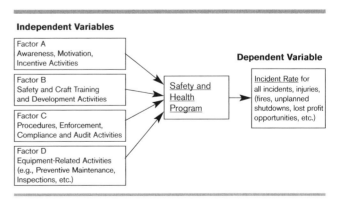

FIGURE 3. Safety and health program model (adapted from Haight et al. 2001a)

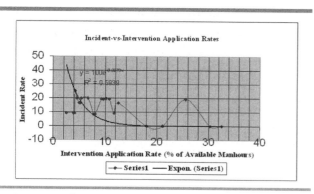

FIGURE 4. Intervention application rate vs. incident-rate relationship (adapted from Haight et al. 2001b)

program set up as an input-output model. This graphic representation shows how one might consider that the program has some quantifiable input and output variables that would allow program effectiveness to be measured and designed to achieve predictable results.

Each of the input variables represents an activity implemented as part of a safety and health program, and each requires the allocation of human resources. How much manpower to allocate and to what extent is not known. The objective is to collect program input and performance data that can be analyzed to develop the mathematical relationship between input and output. Data were collected in the Haight et al. (2001a) study and in the Iyer et al. (2004) study, and a typical mathematical relationship is shown in Figure 4.

The resulting complete mathematical relationship from the Iyer et al. (2004) study is:

$$\begin{aligned}\rho &= \text{Ln}\,[\text{Ln}\,(\hat{y})] \\ &= x(1) \times x(2) \times x(3) \times x(4) - 3.13 \times x(2) \times x(3) \times \\ &\quad x(4) - 1.42 \times x(1) \times x(3) \times x(4) + 0.07 \times x(1) \times \\ &\quad x(2) \times x(4) - 2.34 \times x(1) \times x(2) \times x(3) + 4.79 \times \\ &\quad x(3) \times x(4) - 0.26 \times x(2) \times x(4) + 8.10 \times x(2) \times \\ &\quad x(3) - 0.23 \times x(1) \times x(4) + 2.15 \times x(1) \times x(3) - \\ &\quad 0.096 \times x(1) \times x(2) + 0.76 \times x(4) - 6.94 \times x(3) + \\ &\quad 0.25 \times x(2) + 0.41 \times x(1) - 0.14\end{aligned}$$

This is the best-fit equation representing the mathematical relationship between manpower allocation to specific safety and health program tasks where ρ represents the incident rate, x is the intervention application rate for each specific intervention activity, and the numerical value is the fit factor.

This equation is an example from a specific operation and can be used only for that operation, as the operation is where the equation's data came from. Each operation's equation would look a little different; this one is presented for illustration purposes. One can take each activity in a safety and health program, allocate the indicated percentage of available manpower to each, and predict relatively accurately what the resulting incident rate will be with that allocation of resources (Iyer et al. 2004). This model is still emerging, but the five publications currently in existence on the subject indicate that it is useful, viable, and accurate in choosing which intervention activities should be implemented to prevent injuries and loss incidents and to determine the appropriate level of the activities.

New research has been developed to further the work of Haight et al. 2001 and 2003 and Iyer et al. 2004 and 2005. It provides an advanced analytical method for the development of an effective safety intervention program with the aim of minimizing incident rates (Shakioye and Haight 2010, Oyewole et al. 2010). Over a two-year period, incident-prevention-directed intervention activity data and incident rate data were collected from an American-owned oil company in Nigeria. From these data, an analysis was completed and a mathematical model was developed to allow measurement and optimization of the effectiveness of a suite of interacting intervention activities that minimize incident rates while concurrently allowing for the development of an improved resource allocation plan and strategy. In this work, Oyewole et al. (2010) investigated five main intervention program factors (Factor A: Leadership and Accountability; Factor B: Qualification Selection and Pre-Job; Factor C: Employee Engagement and Planning; Factor D: Work in Progress; Factor E: Evaluation, Measurement and Verification). As in previous work (Haight et al. 2001 and 2003 and Iyer et al. 2004 and 2005), this was done to define the mathematical relationship between model input and output and to show the effects of each factor and their interactions on the incident rate. Oyewole et al. (2010) completed an analysis of variance, which showed that four safety factors (A, C, D, and E) had a significant effect. The innovation that Oyewole et al. (2010) brings to this line of research is the use of response surface design plots to determine the resource allocation method. This particular incident prevention intervention model indicates that the incident rate can be minimized at an optimum allocation of 16.66 percent of the available human resource time. From this, one can determine the significant safety intervention activities that have the most certainty for achieving the desired incident rate.

In order to reap the benefits of this research, it will be important to concentrate more effort and resources on model-indicated significant incident prevention intervention activities that contribute significantly to minimizing incident rates. Using surface response methodology, a point on the model-generated surface can be selected as the point at which the incident rate is at its minimum point. One then finds the human resource time allocation for each intervention activity shown on the graph to use in designing the incident prevention program. One chooses the activities indicated on the graph (the model indicates that they significantly contribute to incident prevention) to determine how much available resource time should be allocated to the indicated incident prevention activities. An example of these results are shown in Figures 5 and 6. From these figures, one can determine the optimum values for each interactive factor from the *minimum* or *lowest* point on the three-dimensional plot (Shakioye and Haight 2010, Oyewole et al. 2010).

This information includes the development of the previous Haight and Iyer work and now provides a more direct solution to the model. These results allow an optimized solution to be chosen from the factors that have been analyzed and presented on the three-dimensional graph. In this case, the results allow a manager to allocate a specific amount of available human resource hours to specific intervention activities from each of the two factors shown. Of course, all models are just mathematical representations of real life, so it is expected that those responsible for allocating resources use the results from these types of models in their own decision making while also incorporating their experience, intuition, and expectations.

Oyewole et al. (2010) further proves that, in order to thoroughly measure the effectiveness of an incident prevention program, one must understand all leading indicators and all lagging indicators (or performance-driving factors) within a loss prevention system and then must define and develop the mathematical relationship between the leading and lagging indicators. It further appears from this work, that it is not enough to measure leading indicators alone or to base resource allocation decisions purely on lagging indicators either. One must determine all performance-influencing factors and build an input-output model around those factors in order to truly understand how to design a loss prevention program and truly drive an incident rate lower. If a suite of intervention activities that make up a loss prevention program are not designed using this methodology, one is leaving this

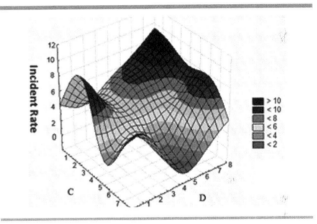

FIGURE 5. Response surface plot of incident rate vs. factors C and D (Oyewole et al. 2010)

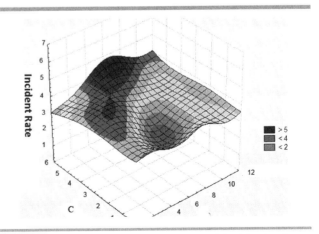

FIGURE 6. Response surface plot of incident rate vs. factors C and D (Oyewole et al. 2010)

important function up to chance. It is suggested that the reader refer to these studies in the reference list to learn more about the process.

Training and Learning Curves

A workforce is one of any company's most valuable and important resources. It is critical that this workforce be skilled and capable of performing its functions in the most efficient, correct, and safe way. To achieve this skilled performance, a management team must provide a mechanism to ensure that the workforce is adequately and appropriately trained. A project manager overseeing a safety-engineering-related project or any type of project involving safety-engineering-related work must incorporate compliance issues as well as skill issues into training programs.

While many training courses are compliance-driven, especially in the safety engineering area, some training and learning needs are not so well established. To effectively address the important training needs of the workforce, one of the first things a project manager or management team should do is to conduct a *training needs assessment* to determine the subject matter that should be addressed in the training program. This needs assessment should center around the specific hazards to which the workers could be exposed, the specific work or tasks they are expected to perform, and specific improvements that need to be made in the operation. The assessment specialist must consider who does what jobs as well as what they do and where, when, and why they do them. This helps the organization to make a distinction between training and nontraining solutions; it helps them to understand problems thoroughly instead of jumping to a training solution before the problem is fully or adequately understood; and it helps to ensure that money and time are not wasted by providing training courses that do not solve the problems of the organization (Hagan, Montgomery, and O'Reilly 2001). Once the training needs are defined, training experts can take over and develop learning objectives, the course content, the lesson plans, and training-frequency requirement schedules for the target audience.

When implementing a training program, care must be taken to ensure that it addresses not only the overall needs of the group of workers but also that it is directed in such a way as to provide for the individual needs of each trainee. Everyone does not learn at the same rate, so project managers must ensure that the training staff properly evaluates the learning curves of a group of trainees for the specific material in which they are being trained. Trainers must first recognize that learning is a time-dependent phenomenon and then address the progress people are making along their particular curve using a process in which the time it takes someone to master a particular skill or task is measured. Depending upon a task's complexity, it could take days or even weeks for people to master a skill to the point where they possess the necessary mental and physical capabilities and coordination to successfully and efficiently transition from one task element to the next without hesitating.

A learning curve is typically exponential and usually indicates that the more people practice a skill, the more often they can perform it quickly, efficiently, or with higher-quality results. If, for example, a person is required to don a self-contained breathing apparatus (SCBA), it usually has to be done very quickly due to the emergency nature of such an action. Once people are taught how to don this equipment, their first attempt at doing it will probably take much longer than their hundredth attempt. The learning curve in Figure 7 shows that the more times people donned breathing apparatus, the better they got at doing it and the faster they did it.

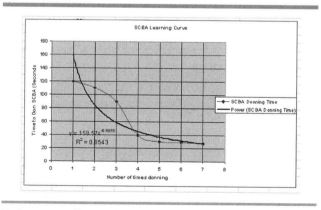

FIGURE 7. Example of a learning curve plot

The learning curve, in this case, is a power curve of the form:

$$y = kx^n \qquad (7)$$

where

y = time to achieve task
x = number of task performances
n = exponent representing the slope of the curve
k = value of the first task attempt

To determine how many times one must don self-contained breathing apparatus before achieving the expected time of 25 seconds, substitute 25 seconds into the equation, take the log of both sides and solve for x:

$y = kx^n$
$25 = 158.57 x^{-0.9058}$
$\log_{10}(25/158.57) = -0.9058 \log_{10} x$
$\log_{10} x = -0.8023 / -0.9058$
$x = 10^{0.8857}$
$x = 7.69$ or 8

The results of this learning-curve analysis indicate that one must practice putting on the SCBA at least eight times to achieve 25 seconds donning time. If the training course does not have a provision to allow for at least eight practice attempts, its objectives must be reevaluated. Whatever the acceptance criteria, one can determine the learning curve in the same fashion. For example, if the task were learning to set up and wear a fall-protection harness system, the trainer could collect data to determine how many practice sessions are needed before the fall-protection system is set up and donned without errors (Niebel and Freivalds 1999).

When developing and managing a training program, it is critical to ensure that if training is being proposed in response to a safety-engineering-related problem, a significant effort is expended to be sure that training is the correct solution to the problem. Training is generally the correct approach if the desired goal is to provide "how-to-do" information to workers. Training is done for the purpose of changing behavior or increasing knowledge, improving performance, reinforcing operational goals, reducing incidents, improving efficiency, or reducing costs. It is not necessarily the solution to a problem in which workers do not do what they are supposed to do. Many times workers take incorrect actions even when they know the correct actions and how to carry them out. Much money is lost each year in training costs because organizations make a decision to train their workforce when a nontraining solution is more appropriate. For example, if workers won't wear their protective equipment because it is heavy or bulky, fogs up, or is otherwise uncomfortable to wear, putting everyone through a 4-hour training session on the importance of wearing protective equipment just to get them to wear it is a waste of time and money. During the needs-assessment stage in the process, a project manager should ensure that all possible training and nontraining solutions are explored (Hagan, Montgomery, and O'Reilly 2001).

Training programs are of many shapes and sizes—they have to be in order to address the many possible subject areas that must be covered. They must also be flexible enough to work around day-to-day production operations and the possibility of high turnover (especially in construction and maintenance turnaround projects, and project work in general). They have to account for individual differences in the workforce, such as differences in experience levels, physical and cognitive capacity, trade-based knowledge, health, age, dexterity, and the types of tasks each worker performs. Training programs must include a provision for ensuring that individual workers understand the training they receive. This can be in the form of a written or oral examination, a hands-on demonstration (for example, an employee shows how to make a safe lift after taking an ergonomics training class), or an assignment for the trainee to now train others (Niebel and Freivalds 1999).

Training can take these many forms and still be effective. It does not have to involve workers sitting in a classroom. It can be as simple as asking employees to read a new procedure or watch a videotape, or it may be all on-the-job training in which employees spend their early years with an experienced mentor. Workers can use simulators that put them in a very real, joblike situation but without the consequences of wrong actions (such as pilots being trained in a flight simulator—there is no crash after an error). Training can involve physical activity such as learning to take flammable gas concentration measurements inside a confined space by actually testing air in a confined space under the guidance of a qualified gas tester (Freivalds and Niebel 2008). Training techniques can involve case studies, facilitated discussions, role playing, lectures, drills, brainstorming, independent study, videos/DVDs, or even computer-aided or online training (Hagan, Montgomery, and O'Reilly 2001).

It addition to the content area of a training program, project managers must consider logistical and administrative issues. Issues such as instructors, location, schedule, frequency, student availability, testing, and record keeping require resolution before beginning any safety engineering training. For more detailed information on training, please refer to the "Safety and Health Training" section in this handbook.

Work Sampling

More traditional allocation methods can be relied on when tasks are well defined. *Work sampling* is a method of observing a task while it is being performed to determine whether a specific outcome will occur. It is described somewhat by the laws of probability. The probability that event x will occur in n observations is

$$(p + q)^n = 1 \qquad (8)$$

where

p = the probability of a single occurrence
q = $(1 - p)$ (the probability of an absence of the occurrence)
n = the number of observations

The distribution of many of these probabilities follows binomial distribution laws. It is known that as n becomes larger, the binomial distribution begins to approach a normal distribution. Therefore, with large enough numbers of observations, one can use the normal distribution laws to approximate the results of a task. Since this is the case, p can represent the mean of the distribution and the standard deviation of $\sqrt{pq/n}$ is approximately a normally distributed random variable. Using the laws of normal distribution, we can estimate that 95 percent of the expected outcomes of our observations will fall within two standard deviations of the mean. For a binomial distribution, the function is as follows (Niebel and Freivalds 1999):

$$F_x = (x; p, n) = P(x) = (^n{}_x) \, p^x q^{n-2} \qquad (9)$$

where

n = the number of samples
x = the number of successful samples
p = the probability of success
q = the probability of failure

QUESTION FOR STUDY

You are managing a job involving a confined-space entry and are worried about possible delays caused by problems with the availability of supplied air. You can get an idea of both the probability of the unavailability occurrence and the approximate delay time by using this equation. Observations over the previous year indicate that supplied breathing air was unavailable 8 percent of the time, and 25 confined-space-entry jobs were performed. This caused delays to the projects totaling 10 days (no entry work can be done without breathing air). Twenty-five more confined-space-entry jobs are planned for 2012, and all require the use of supplied breathing air. What is the probability that two jobs will be delayed due to unavailability of supplied breathing air?

$n = 25$
$p = 0.08$
$q = 0.92$
$x = 2$

$P(x = 2) = (^{25}{}_2) \, p^x q^{n-2}$
$P(x = 2) = (300)(0.08)^2(0.92)^{23}$
$P(x = 2) = 0.282$

With a probability this high, you might consider adding back-up respiratory equipment or more people.

Another example of work sampling with a quantitative sampling analysis method uses a Poisson distribution. The Poisson distribution determines discrete probabilities of a specific number of events occurring in a known amount of time if there are adequate observation data and those data follow a Poisson distribution (Eisner 2002):

$$P(k) = (\lambda t)^k e^{(-\lambda t)}/k! \qquad (10)$$

where

$P(k)$ = the probability of exactly k events of interest occurring
λ = the rate at which the events occur
t = the time over which the events occur

When observing and sampling any type of work for the purpose of making risk decisions, or even just likelihood decisions, the appropriate number of observations or data points needed to make statistically supported decisions must be determined. To accomplish this, the desired accuracy requirements must be identified. The basis for this consideration is that the more observations made, the more valid the final answer will be from the standpoint of predicting an actual outcome. To do this, the following formula can be used:

$$n = 4p(1-p)/l^2 \qquad (11)$$

where

n = the number of observations
p = the probability of occurrence of the event you are trying to prevent
l = the error limit or desired accuracy

QUESTION FOR STUDY

Equation 10 could be used to analyze a situation in which you want to know how many hot-work and confined-space permits you will be able to process in one hour given the current manpower availability.

In the process of planning a major maintenance turnaround, you are concerned that the necessary day-to-day work permits required to keep the turnaround on schedule will be too high, given the number of employees assigned, to evaluate and approve safety-related permits. Bid estimates require that at least ten permits be processed in an hour. This appears to be a potential bottleneck for the project, so past work-sampling data are evaluated and a Poisson calculation is undertaken to determine the probability that five permits can be processed every 30 minutes (given the current staffing and availability of analysis equipment). Over the last four maintenance turnarounds in this unit, permits have been processed at an overall rate of eight permits per hour. The probability that five permits can be processed in 30 minutes (or $t = 0.5$ hours) is:

$P(5) = [(8)(0.5)]^5 e^{[-(8)(0.5)]}/5!$
$P(5) = (1024)(0.0183)/120$
$P(5) = 0.156$

This probability is low and indicates that a project slowdown is likely to occur. If this is a critical turnaround and the unit has to be back up and running by exactly the originally scheduled turnaround completion date, this probability—15.6 percent that the project team can process permits at the rate needed to stay on schedule—means that staffing allocations will have to be reevaluated.

QUESTION FOR STUDY

You want to determine the likelihood that employees are not wearing proper protective equipment. It has been reported that 10 percent of the employees are not wearing their protective equipment, but you claim that the violators make up only 5 percent of the employees (Niebel and Freivalds 1999). To confirm this, you want your accuracy to be plus or minus 2 percent. The number of observations needed is

$n = 4(0.05)(1 - 0.05)/(0.02)^2$
$n = 475$ observations

If you can afford the time to make only 250 observations, you can solve for the expected accuracy by:

$$l = \sqrt{4(p)(1-p)/n} \qquad (12)$$
$l = \sqrt{4(0.05)(0.95)/250} = 0.0275$

or ±2.75% accurate

In general, many types of analyses can be done to understand the work going on in an operation. Safety engineering project managers must first understand the performance output they are most interested in and then determine whether successful performance is related to time (on schedule), number of occurrences (number of permits processed), or number of people required to safely complete a job. Once the desired output and its expected quality are known, they need to determine and quantify the variables that affect the output performance. The rest of the work-sampling approach involves data collection (performance variables) and analysis (the appropriate analysis for the resulting distributions). This sampling and analytical process will aid decision making and help project managers to ensure an efficient operation that completes work on schedule and within budgetary constraints, while still allowing for safe completion.

Performance Ratings

There is no universal method to rate worker performance in terms of productivity, efficiency, safety, or accuracy. There is no universal measure of what represents normal performance. For each task in a safety engineering project, acceptable performance must be defined. Consideration must be given to what task completion comprises, how much time it should take to carry out each task, and how acceptable quality is determined.

Benchmarking is often used to help determine these characteristics of a task. Standards and acceptance criteria are established for each step in a task from benchmarking observations of the same task being completed at another company location or at a different company's operation. From the acceptance criteria, a performance rater can be trained to do the observation (he or she will know what to look for) and an actual performance rating can be carried out.

When benchmarking, determine whether all necessary steps (as defined by initial internal task design) are included. If not, the benchmarking analyst should ascertain the reasons the steps are missing and resolve the situation by redesigning the task or deciding that the missing steps are not necessary. The analyst should determine what equipment is being used for the job and decide whether it is appropriate. He or she should determine how long each step in the task takes to complete and how to determine when it is complete. Each of these criteria should be documented in a form that is easy to follow for a performance rating analyst observing the same task in the future.

The acceptance criteria must take into account the fact that there are individual differences in the people carrying out the task. Differences in general fitness, experience, dexterity, strength, age, level of training, and so on may affect any or all aspects of task performance. The analyst who develops the performance rating criteria must define acceptance in terms of a range that takes into account the fact that workers should be well adapted to the task, adequately trained and experienced to perform the task, in adequate physical shape, and coordinated both physically and mentally to a level that allows efficient movement from one step in the task to the next, while obeying the principles of motion economy (Niebel and Freivalds 1999, Freivalds and Niebel 2008). The analyst may consider recording time-and-motion data for all steps and developing performance ranges and frequency distributions that allow the establishment of a statistically supported acceptance range.

For example, this performance rating may be applied to an excavation job because each of the preparation, permitting, and implementation steps can be observed and rated against a known acceptance standard. The project person responsible gathers all relevant subsurface soil data, identifies the location of the excavation, seeks appropriate concurrence and approvals for the job, seeks appropriate approval signatures on the permit, arranges for appropriate equipment and manpower, ensures appropriate excavation design (e.g., step-back method or shoring), ensures appropriate dewatering of excavation and deposition of the spoil pile, and so on. Each of these steps receives a score based on whether the steps taken match established excavation and permitting protocol. A performance rating analyst scores each task as present or absent—and if present, to what quality is it being applied—and overall time it takes to complete the job. An indexing score can be used to allow consistent relative ranking on quality.

This process is the same as ones used by many organizations implementing behavior-based safety observation programs. These programs are, in essence, performance-rating processes. In the case of behavior-based safety programs, the ratings are based on an observer-developed, critical, acceptable behaviors library. All observations are done by comparing actual performance to the critical behaviors definition. Task performance is rated either *safe* or *at risk*. At risk, or unacceptable task performance is addressed and changed so that a *safe* score can be achieved.

Even though there is no universally accepted standard, after a task or a project is observed long enough, a picture emerges of what it should look like. The challenge then becomes to capture that acceptable performance on an observer-used data-gathering tool, such as a checklist, a rating sheet, or a scoresheet. These ratings can then be used to address individuals or contract companies whose performance does not meet expected, accepted levels.

Time Study and Allowances

The concept of *time study* has been a part of industrial engineering for many years. It is a process by which an observer watches a particular task being performed many times by many people; from the observation data, time standards are developed for each target task. Its main purpose is to see that a task is carried out efficiently and with minimal wasted effort, motion, or time. The method is based on the measurement of the content of a task based on a specific and defined method for carrying out the task, given that there are adequate allowances for fatigue and for certain unavoidable delays. Adequate data should be collected to ensure that the time standards established for a particular task are in fact accurate and are also based on sound statistical significance.

The first step in this process is to divide a job into its component elements. Once this is done and the task steps for each job are established, the dispatch of each component task is timed and recorded. Once a statistically significant number of observations has been obtained and appropriate allowances have been considered, a suitable time standard is established and used for all future time-study observations. Allowances that must be considered in establishing time standards are fatigue, abnormal posture requirements, muscular force application requirements, illumination levels, atmospheric conditions (e.g., too hot, too cold, damp, icy, slippery), noise levels, visual strain, mental strain, monotony and tediousness, as well as a constant allowance for personal needs (work stoppage related to maintaining personal well-being).

These allowances were quantified by the International Labor Office in 1957 and are still used today for the most part (Niebel and Freivalds 1999, Freivalds and Niebel 2008). Allowances are considered after normal time (NT) has been determined by the formula NT = OT (observed time) × $R/100$, where R is a rating based on a percentage of efficiency compared to 100 percent—a level applied by a qualified, experienced operator working under normal conditions at the workstation with no undue time burden or pressure. Knowing normal time and allowances, standard time for each task can be determined. The formula for determining the standard time (ST) for completing a job is:

$$ST = NT + NT \times \text{allowance} \qquad (13)$$
$$ST = NT \times (1 + \text{allowance})$$

For example, if a task step in a job has a normal time of 0.145 minutes and a particularly high level of mental strain is associated with it, requiring an allowance of 0.08 (considered "very complex" in the International Labor Office, 1957 recommendations), the standard time would be:

$$ST = 0.145 \text{ min} \times (1 + 0.08)$$

An ST of 0.156 minutes should be allowed to complete this task.

Time-study and allowances consideration is quite involved and often for safety-engineering-related work is not necessary; however, someone charged with managing safety-engineering-related work should be familiar with time study and should understand thoroughly the concept of allowances. Many allowances are necessary to preserve the health and safety of workers. Therefore, safety engineering managers are encouraged to do additional reading on the subject of time study and allowances, as can be found in Niebel and Freivalds (1999).

PROJECT COMPLETION
What does completion mean?

In day-to-day operations, it is often the case that safety engineering work is never complete. Everyday operations require the input of the safety engineer, and, in fact, the safety engineer often assists line management personnel in managing their safety-related responsibilities. *Completion* as discussed here refers to finite project work in which projects such as construction, installation, and maintenance are undertaken, managed throughout their course, and then closed out as the effort is completed and the workers are reassigned. When the building construction is completed, the pipeline is laid, or the paper machine repair work is completed, the project is finished and it is time to assess performance. This closeout work may mean that scaffolding is taken down, excavations are covered, open vessels are resealed, and the workers depart. Project records must be addressed and information worked into the plans for the next time the project is undertaken. Lessons learned from incidents, injuries, or other problems must be sorted out and communicated. There is much to be done as and after a project is completed.

For this effort, one should rely on an element from the field of controlling referred to as *post-action* or *feedback controls*. While active pre-action (feed-forward) and concurrent (steering) controls have been in place throughout the project to help manage the project itself, the concept and tools of feedback or post-action controls are needed in order to learn from project results and improve them the next time such a project is implemented (Schermerhorn 1993).

One example of a post-action control is the final project budget summary. It will not help project managers to get the budget under control, but it will help to determine where they should place emphasis next time if there is a budget overrun. Another example is a final site inspection. This post-action control may tell a project manager that the project is not really finished. If a state regulatory agency does a post-clean-up inspection after an asbestos removal job and finds high concentrations of asbestos in the air, the project team will likely have to come back and continue the removal and clean up. Post-action controls such as these are intended to help project managers (especially those managing the safety engineering aspects of a project) in the planning process and to improve project performance the next time the work is done. They can also be used to document and recognize good performance during the current project (Schermerhorn 1993).

Project Closeout

When project work is completed, proper closeout is critical for many reasons, including financial concerns, but most importantly from a safety engineering point of view. In many cases, the project management system includes a provision requiring a closeout inspection of the work. One Occupational Safety and Health regulation (29 CFR 1910.119, *Process Safety Management of Highly Hazardous Chemicals*) calls this step the "pre-startup safety review." It is critical that buildings are not occupied, vessels are not pressured, conveyor belts are not started, flammable liquids are not introduced, and earth is not tread upon until there has been some type of inspection to determine that all is structurally sound, all codes have been met, all affected workers have been trained, and all facility documentation (e.g., maps, plot plans, piping and instrument diagrams, procedures) have been revised and updated. A number of projects over the years illustrate the importance of such a closeout phase in a project—for example, wrenches and hard hats (left in a vessel) clogged flowing liquid systems and parking decks collapsed. There was even a situation in which someone wanting to do a final inspection inside a vessel didn't follow the confined-space-entry rules and was nearly closed up in the vessel, which was about to have raw materials, heat, and pressure reintroduced (this last example is from the author's own history).

Many disciplines should be involved in the closeout process. Project managers must make arrangements to include representatives from all of the required disciplines, including regulatory agency representatives who may be required to issue such documents as certificates of occupancy.

Operating experts are required to ensure that the system and equipment can be operated by the people who will need to operate it.

Accountants and cost analysts are required to track the budget performance.

Planning (time and resource allocation) analysts may be required to determine schedules and inventory performance.

Structural engineers and safety engineers may be required to determine whether the buildings, ladders, stairwells, decks, flooring, earthen dikes, accessways, support structures, and so on were built as designed and are adequate for their intended purpose.

Electrical and instrument engineers may be required to determine whether the wiring, power generating and supply, and control systems are built to code, as designed, and are adequate for the intended service.

Fire protection engineers may be required to determine whether exits, exit accessways, fire suppression systems, detection and alarm systems, fire department access, and water supplies are adequate and meet required codes and that emergency response provisions have been met.

Environmental engineers may be required to ensure that pollution prevention systems have been installed properly and will perform as intended.

Safety engineers may be required to determine whether the site has been adequately cleared of all equipment, parts, and supplies that could make access to and travel through the site unsafe. They may also be required to determine and ensure that documentation has been adequately and thoroughly revised to reflect the new facility or system.

Regulatory agency experts or officers may be required to do an overall project or site inspection to ensure adequate clean up (asbestos), sizing of storm water drainage (water pollution), adequate incinerator stack height (air pollution), and so on.

The records of these inspection results must be maintained and an approval signature must be received for each. Project managers should not allow start-up and use of the new facility until they have been adequately assured by each expert that it is safe and approved to do so. While experts in each discipline may have their own systems, checklists, or similar means to determine whether the facility is safe to start up will help to ensure consistency. Safety engineers may assist in the project by developing, in collaboration with a representative of each discipline, these inspection checklists. The checklists not only facilitate the inspection, but they can also form the record of the inspection. If a provision is made for an approval signature on the checklist, everything is in one package (Hagan, Montgomery, and O'Reilly 2001). It stands to reason that after all representatives with a vested interest in the project and the expertise to evaluate it have given their approval, it is safe to initiate occupancy and/or operation.

Presentation of Results

While project work is going on, presentations of ongoing status information are likely to be given on a regular and frequent basis. Usually these presentations are less formal and contain less information as well as less-detailed information than a final presentation of all the project results. These presentations serve a somewhat different purpose than the presentation of the final project results in that their objective is usually to help people to know where they need to steer things back on track. Since they are usually less formal, they may take many formats, so they will not be treated here. This section addresses the larger, more formal, final presentation of project completion results.

The purpose of presenting final project results is similar to the objectives discussed above for feedback or post-action controls. It is desirable to know how the methods employed during the project worked so that the results can be used in planning for the next project. It is also desirable to use this forum to recognize and document positive accomplishment (Schermerhorn 1993). If outside contractors are involved in a project and performance incentive clauses are built into the contract, this presentation also recognizes the achievement of performance awards.

Some issues that can or should be addressed in a final project presentation are budget and costing performance; completion schedule performance; safety performance; equipment condition reports (for maintenance turnarounds); new building, facility, or system uses; output improvements (production capacity increases); and new products to be manufactured or new markets to be entered. It may be appropriate for this presentation to include some discussion of the ongoing needs or responsibilities of the new or newly renovated facility.

This presentation should be developed, managed, and given like any other presentation in that it should be guided by the same established rules of effective presentations that guide other presentations. Presenters should know the audience and what it will be most interested to hear. They should tell the audience what they are *going* to tell them, what they *are* telling them, and then what they *told* them. They should present material visually (e.g., PowerPoint-type slide) and explain what the visual material means, but not read from the visual material. They should make sure that the visual material is not too busy; the information can be seen or read by everyone in the room (considering lighting, clarity, contrast, and so on); the information is not condescending to an audience that already knows the subject matter; and the slides are not too flashy, colorful, or otherwise distracting. During the presentation itself, presenters should make eye contact with everyone and ensure that the pace allows people some time to think about each point being made. They should decide how to handle questions (during or after the presentation) and provide hard copies of the presentation material after it (Eisner 2002).

Presenters also must make some decisions about the presentation format and logistics. Considerations include whether the content will be lecture-based or activity-based, the number of speakers and the logistics associated with changing speakers, the topics and information that will be covered (just a bottom-line status report of costs, duration, and incidents, or a discussion of lessons learned), and whether there will be roundtable open discussion or a closed presentation of the facts. They must consider lighting (e.g., no bright lights over the projection screen, adequate lighting for attendees to take notes), break schedules (attendees may need a break shortly after lunch as well as at every hour)

and getting attendees to return from breaks on time, comfortable seating (but not so comfortable as to induce sleep), room temperature (public spaces are often kept so cold that participants cannot focus on presentations, so presenters should ensure that the room is not too cold), refreshments, cell phone use, and laptop use (and availability of power for them). Even though there is no correct combination of decisions and conditions (this is very site- and situation-specific), a good presentation given under the right conditions will often be a successful presentation (Eisner 2002).

Ongoing Implementation, Record Keeping, and Communication of Lessons Learned

When the final presentations are completed, an organization will be best served by an effort in the routine management plan to capture and make use of all that was learned during the implementation of the project. The objective is to ensure a successful implementation the next time a similar project is undertaken. Since people change jobs and move on to other issues, a mechanism and a system must be in place to record, retrieve, and communicate the valuable lessons learned and the important records from the project.

The information is the same as that discussed in the final presentation of project results. The cost and budget information will be useful for future estimating purposes. The schedule performance information, broken down by task or job segment, will also contribute to manpower and time estimation for the next time. Materials of construction, repair details, spare-part usage rates, equipment condition inspection records, protective equipment availability and usage rates will all help in putting the right people and equipment on the job in adequate levels and quantities. Training topic and attendance records, meeting topic and attendance records, contractor turnover records, injury rates, and industrial hygiene monitoring program and results will help to determine what the safety and health program should look like for the next project. Information on traffic patterns, laydown areas, delivery locations, and equipment storage issues will help to relieve congestion and ensure that adequate space is allotted for implementation of similar projects in the future.

In general, all of these records, lessons, and information should be maintained and used in the planning process of a new project. They should be communicated to the planning team set up to implement the next project to ensure that old lessons are not forgotten and new people on the project do not waste time solving problems long since solved.

Conclusion

In managing any project, in the safety engineering area or any other area, one must consider the budget, the schedule, protection of the workforce, and the objectives of the customer, as well as learning something from the mistakes made. While managerial skills are critical, people manage this type of work through effective leadership. They have to rely on strong interpersonal skills, all of which may not be teachable but are acquired over time through living and experience. All the analysis in the world may not guarantee the successful completion of a project. We all have to remember that managing work means managing people, and managing people means relying a lot on what we learned in kindergarten and the first grade.

References

Accreditation Board of Engineering and Technology (ABET), Inc. *Criteria for Accrediting Applied Science Programs, 2002–2003 Accreditation Cycle*. Baltimore, MD: ABET.

Brauer, R. L. 2005. *Safety and Health for Engineers*. New York: Van Nostrand Reinhold.

CoVan, J. 1995. *Safety Engineering*. New York: John Wiley & Sons, Inc.

Duderstadt, J. J., G. F. Knoll, and G. S. Springer. 1982. *Principles of Engineering*. New York: John Wiley & Sons, Inc.

Eisner, H. 2002. *Essentials of Project and Systems Engineering Management*. 2d ed. New York: Wiley Interscience, John Wiley and Sons, Inc.

Freivalds, A., and B. Niebel. 2008. *Methods, Standards and Work Design*. 11th ed. Boston: WCB McGraw Hill.

Gloss, D. S., and M. G. Wardle. 1984. *Introduction to Safety Engineering*. New York: John Wiley & Sons.

Iyer, P. S., J. M. Haight, E. del Castillo, B. W. Tink, and P. W. Hawkins. 2004. "Intervention Effectiveness Research: Understanding and Optimizing Industrial Safety Programs Using Leading Indicators." *Chemical Health and Safety American Chemical Society Division of Chemical Health and Safety* 11(2):9–19.

_____. 2005. "A Research Model—Forecasting Incident Rates from Optimized Safety Program Intervention Strategies." *Journal of Safety Research* 36(4):341–351.

Hagan, P. E., J. F. Montgomery, and J. T. O'Reilly. 2001. *Accident Prevention Manual for Business and Industry – Administration and Programs*. 12th ed. Itasca, IL: National Safety Council.

Haight, J. M., and R. E. Thomas. 2003. "Intervention Effectiveness Research—A Review of the Literature on 'Leading Indicators.'" *Chemical Health and Safety—American Chemical Society—Division of Chemical Health and Safety* 10(2):21–25.

Haight, J. M., R. E. Thomas, Leo A. Smith, R. L. Bulfin, and B. L. Hopkins. 2001a. "Evaluating the Effectiveness of Loss Prevention Interventions: Developing the Mathematical Relationship Between Interventions and Incident Rates for the Design of a Loss Prevention System (Phase 1)." *Professional Safety— The Journal of the American Society of Safety Engineers* 46(5):38–44.

_____. 2001b. "An Analysis of the Effectiveness of Loss Prevention Interventions: Design, Optimization, and Verification of the Loss Prevention System and Analysis Model (Phase 2)." *Professional Safety—The Journal of the American Society of Safety Engineers* 46(6):33–37.

Hersey, P., and K. Blanchard. 1977. *Management of Organizational Behavior Utilizing Group Resources*. 3d ed. Englewood Cliffs, NJ: Prentice Hall.

Lack, R. W. 1996. *Essentials of Safety and Health Management*. Boca Raton, FL: Lewis Publishers, CRC Press LLC.

Louis, M. R., and R. Sutton. 1991. "Switching Cognitive Gears: From Habits of Mind to Active Thinking." *Human Relations* 44(1):55–76.

Nahmias, S. 1993. *Production and Operations Analysis*. 2d ed. Homewood, IL and Boston, MA: Richard D. Irwin, Inc.

Niebel, B., and A. Frievalds. 1999. *Methods, Standards and Work Design*. 10th ed. Boston: WCB McGraw Hill.

Occupational Safety and Health Administration (OSHA). 29 CFR 1910.119. (accessed 15 January 2005). www.osha.gov

Oyewole, S. A., J. M. Haight, A. Freivalds, D. J. Cannon, and L. Rothrock. 2010. "Statistical Evaluation of Safety Intervention Effectiveness and Optimization of Resource Allocation." *Journal of Loss Prevention in the Process Industries* 23(5):585–593.

Park, C. S., 1997. *Contemporary Engineering Economics*. 2nd ed. Menlo Park, CA: Addison-Wesley Longman, Inc.

Schermerhorn, J. R. 1993. *Management for Productivity*. 4th ed. New York: John Wiley and Sons, Inc.

Shakioye, S., and J. M. Haight. 2009. "Modeling Using Dynamic Variables—An Approach for the Design of Loss Prevention Programs." *Safety Sciences* 48(1):46–53.

Wright, P. H., A. Koblasz, and W. E. Sayle II. 1989. *Introduction to Engineering*. New York: John Wiley & Sons.

GLOBAL ISSUES

4

Kathy A. Seabrook

LEARNING OBJECTIVES

- Be able to demonstrate through examples how worker safety and health risk impact the economic and operational integrity of a global company.

- Identify and understand the challenges a multinational company faces when managing global workplace safety and health risks.

- Develop global strategies to manage global workplace safety and health risks.

- Understand how country and business cultures impact global worker safety and health implementation and performance.

- Identify and assess personal workplace risk when working and traveling globally.

THIS CHAPTER provides a business perspective on managing worker safety and health risks as they relate to global companies. Global companies are changing rapidly, whether through joint ventures, organic expansion, acquisitions, or divestiture. These changes, tied to the increased interdependency of global manufacturing and distribution systems, combined with the interconnectivity of global communication, media, and information systems, are significantly impacting global trade and presenting a myriad of new challenges for business (Friedman 2005). One of these challenges is managing worker safety and health risks throughout a global organization.

Using a business-perspective and model-centric approach, rather than strictly a safety and health regulatory-driven perspective, this chapter examines worker safety and health issues as a critical part of the overall strategic planning process companies engage in when running a global organization. It will demonstrate how managing worker safety and health as a business risk can either positively or negatively impact profit, market share, brand, and manufacturing and distribution within international organizations. Whether mitigating reputational risk, complying with safety and health laws, or handling risk litigation, managing worker safety and health is a good strategy for global companies. This strategy ensures that global companies are not prohibited or inhibited from operating in any market, including new or emerging markets.

The role of a safety and health professional within a global company is to provide senior management with both the short- and long-term perspective on how worker safety and health risk impacts the functional, financial, and operational future of the organization. It is therefore critical that the management of worker safety and health risk is embedded in the fabric of global companies as they continue to seek business opportunities in new, emerging, and existing markets.

A Global Business Perspective

The global business community continues to expand into new and emerging markets in an effort to grow their global market share, enhance their profits, and expand their customer base. This, along with the rise in global interconnectivity (instantaneous communication, information sharing, and media access), has changed how global and international companies operate over the past twenty years. For example, access to portals of information, official and unofficial, has changed many global companies' perspectives on what constitutes a risk to their business operations. They are now recognizing that worker safety and health risks can pose as great a reputational risk as the traditionally recognized reputational risks associated with branding, manufacturing operations, marketing, sales, and the financial areas of an organization (BP 2005, Knight 2004).

This globalization trend continues with more competition from private local companies in developing markets. China is an example. According to the National Bureau of Statistics of China, Chinese companies have increased their market share from 3 to 23 percent, while foreign multinationals have only grown their market share from 25 to 31 percent (1999–2009). This domestic growth provides Chinese companies the financial ability, in industries such as mining, to be more aggressive in penetrating markets internationally (Moody 2009).

Integral to globalization is the expansion of manufacturing and distribution systems to meet the organizational growth targets of emerging markets such as China, Brazil, and India. This increases the interdependence of manufacturing and distribution systems in regional and local markets to deliver goods and services. As an example, China has developed export processing zones (EPZs), where multinational companies and joint ventures are finding incentives to locate manufacturing and distribution facilities, enabling them to expand their market share by bringing their products and services to the region (China Association of Development Zones 2002). The State Council in China has developed enclosed management schemes over these EPZs to further open its market to the outside world and attract international business. EPZs are special enclosed areas supervised by the customs authority, and include the cities of Guangzhou, Beijing, and Dalian (China Association of Development Zones 2002).

With the expansion of manufacturing and distribution systems comes the need for a skilled indigenous workforce to manage and operate those systems. In most cases, the cost for educating the workforce is borne by the global company, and competition for these educated, skilled workers is fierce. This is especially true in China. Globalization has both risks and rewards. From a regulatory standpoint, compliance with a country's local safety and health regulations, as well as being a good corporate citizen, is an organization's license to do business in that country, and may very well be a measure of the business' survival.

The remainder of this section demonstrates, through real-world business cases, that worker safety and health risks can and do impact the financial health and future of a global company. These real-world examples clearly demonstrate the need for companies to manage worker safety and health risks as a strategic imperative rather than a simple regulatory compliance exercise.

Reputational Risk

In the following case studies, the significance of reputational risk is demonstrated. Global companies such as British Petroleum p.l.c. (BP) and Nike, Inc. recognize that worker safety and health risks are a leadership issue and have experienced the direct impact of worker safety and health-related incidents on their company's reputation (Nike 2003, BP 2005).

The case studies of BP and Nike, Inc. demonstrate how important it is for global companies to recognize and manage their worker safety and health risks as a strategic business risk that has an impact on the reputational, financial, and operational integrity of their businesses. Figure 1 depicts the interrelationship of these risks.

The BP, Nike, and Union Carbide examples are good reminders that worker safety and health risks must be identified and then managed through an organization's strategic planning process.

CASE STUDY

British Petroleum

British Petroleum (BP) is one of the world's largest global energy companies. An incident on March 23, 2005 at BP Products North America, Inc., in Texas City, Texas, demonstrated to BP's senior management the direct financial and reputational impact of a worker safety and health incident on their company. According to a May 2005 press release issued by BP Products North America, Inc., an explosion in the Texas City facility's isomerization process unit killed fifteen and seriously injured 170 workers at the facility (British Petroleum p.l.c. 2005a).

Immediately following the reports of the incident, BP's senior leadership recognized the reputational risk implications that media coverage of the incident was having on the company. The company, led by John Browne, Group Chief Executive, set out to address the situation and manage all information associated with the incident. According to a communiqué from Browne that was posted on their Web site in May 2005, BP wanted transparency and access to factual information by all stakeholders, including the injured workers and their families (British Petroleum p.l.c. 2005).

BP initiated an immediate investigation of the incident with subsequent implementation of recommendations, along with claiming responsibility for the incident. In a press release, Ross Pillari, who was then President of BP Products North America, Inc., publicly apologized for BP's mistakes. He also disclosed the failure of BP's isomerization unit process managers to provide leadership, by not always being on site during critical periods or verifying that the unit operators were using correct procedures (BP Products North America, Inc. 2005). Subsequently, BP entered into a settlement agreement with OSHA in March 2005 and paid $21,361,500 in fines for violations relating to the incident (U.S. Department of Labor 2009a, 2009b).

March 2005, BP appointed former U.S. Secretary James Baker III, to head a U.S. Refineries Independent Safety Review Panel. The panel was charged with conducting an independent assessment of the company's U.S. refineries and of the company's corporate safety culture (Allars 2007). The resulting report, known as the "Baker Report," was published on January 16, 2007, and found process safety performance problems within BP's U.S. refineries (Allars 2007). In March 2007, the U.S. Chemical Safety and Hazard Investigation Board (CSB) released a report of their investigation, which also found "systemic process safety issues at the BP refineries in the United States" (U.S. Department of Labor 2009b).

According to the OSHA fact sheet on the BP Monitoring Inspection, the September 22, 2009 BP deadline for abatement outlined in the 2005 Settlement Agreement was not met and OSHA issued a "Notification of Failure to Abate" and additional willful citations. The proposed penalties were a record-breaking $87,430,000 (U.S. Department of Labor 2009b and 2009c). BP has contested the fines before the Occupational Safety and Health Review Commission; the Occupational Safety and Health Administration and BP were in settlement discussions as this book went to press.

The events of the Deepwater Horizon drill rig incident were unfolding as this book went to press. According to U.S. President Barack Obama in his weekly address, ". . . BP's Deepwater Horizon drilling rig exploded off Louisiana's coast, killing 11 people and rupturing an underwater pipe. The resulting oil spill has not only dealt an economic blow to Americans across the Gulf Coast, it also represents an environmental disaster" (The White House 2010).

CASE STUDY

Union Carbide

All risks to business are interdependent, whether worker safety and health, operational, financial, personal (senior management), or reputational. Union Carbide experienced this fact in a December 3, 1984 incident at their facility in Bhopal, India. According to the Indian state government of Madhya Pradesh and Union Carbide, the cause of the incident was the introduction of a large volume of water into a methyl isocyanate (MIC) tank, triggering a reaction that led to a gas release. The incident resulted in the loss of approximately 3800 lives, with another 40 individuals experiencing permanent total disability, and 2680 experiencing permanent partial disability, according to the state government for Madhya Pradesh (Union Carbide Corp. 2004).

This worker safety and health incident impacted the company's operations, financials, and reputation, along with raising the question of the personal and criminal liability of the chairman at that time, Mr. Warren Anderson. The incident resulted in litigation, financial settlements, and legal costs. In the end, Union Carbide sold their Indian operations, and today Union Carbide Corporation is a wholly owned subsidiary of The Dow Chemical Company (Union Carbide Corp. 2004).

A more detailed recap of the Union Carbide Bhopal incident and its impact on the company is found in Appendix B of this chapter.

CASE STUDY

Nike, Inc.

Nike, Inc. is the most dominant global athletic shoe and apparel company in the world today (Locke 2005). This place in the global market has come by way of a bumpy road through the hazards of globalization.

In the mid-nineties, Nike faced a series of allegations involving their use of underpaid workers in Indonesia, child labor in Cambodia and Pakistan, and poor (health and safety) working conditions in Vietnam. These workers were not Nike employees, but employees of independent contractors and vendors working within Nike's global sourcing network (Locke 2005). Although these manufacturing sites were not owned or operated by Nike, the company's stakeholders were holding them accountable for their supply chain. This, coupled with the negative media coverage, significantly impacted Nike's reputation and financial performance. At the time, Nike did not recognize the full implications of the reputational risk of this incident. Lacking proactive media management and a coherent response to these allegations, Nike's global reputation and future sales were significantly impacted (Locke 2005).

Nike senior management recognized the growing threat to its brand, its most important asset, and went on the offensive (Gordon 2001). By demonstrating their ability to proactively manage media allegations, and taking responsibility for the actions of their supply-chain network, Nike was once again considered a good global corporate citizen.

With the Nike brand intact, the company launched several socially responsible initiatives, such as the Management of Environment, Safety, and Health (MESH) program and the Global Alliance for Workers and Communities (Nike 2003). MESH was launched to assist Nike's Asian business partners (supply-chain companies) with developing goals, targets, objectives, monitoring systems, and self-assessments for environmental safety and health (ES&H) risks. In addition to supply-chain ES&H management, the program incorporated better manufacturing practices, along with community affairs and health and nutrition programs. Nike also provided educational forums for their contract footwear manufacturers in the development, implementation, and monitoring of their MESH programs (Nike 2003).

Nike also became a partner of the Fair Labor Association (FLA). FLA provides independent assessments of the working conditions of apparel and footwear facilities throughout the world. They publish an annual public report of organizations that have been independently assessed by FLA (Nike 2005a). Nike's involvement directly reflects an understanding of reputational risk and the need to manage workplace risk, including safety and health issues (Nike 2003).

Did Nike, Inc.'s ES&H strategy work to mitigate negative publicity and rebuild their stakeholders confidence? The answer appears to be yes. According to their 2004 annual report, Nike, Inc.'s revenues were just over $12 billion for the fiscal year ending May 31, 2004, a 15 percent increase over revenues for the previous year (Knight 2004). They also capitalized on their international presence, launching Nike-Russia alongside already existing Nike-Brazil, Nike-India, and Nike-China, three of the largest and fastest-growing consumer markets in the world, according to founder and first Nike CEO, Philip Knight (Knight 2004). Nike's disclosures in their subsequent annual reports still raise issues, but the transparency in reporting the management of those issues has had a positive impact on their stakeholders.

The remainder of the sections in this chapter provides an overview of tactical challenges and strategies for managing worker safety and health risks in the global arena. In the author's opinion, strategies are typically considered best practice and are used by global companies to successfully manage their organization's worker safety and health risks on a global level.

CHALLENGES AND STRATEGIES OF GLOBAL WORKPLACE SAFETY AND HEALTH MANAGEMENT

There are significant challenges to managing the worker safety and health risks of a global and international organization. They include leadership commitment, identifying and prioritizing risks on a global level, resource allocation, developing a global management system or process, conformity assessment, developing key global performance indicators, and competent global safety and health resource networks. In addition, one cannot ignore the important challenge of managing the safety professional's own workplace risk while traveling and working outside the home country.

Without a well-thought-out strategy, leadership commitment, and allocated resources, this can sound daunting. What is needed is an understanding of the challenges to expect and the strategies that have proven successful in managing worker safety and health risk throughout an organization's global operations.

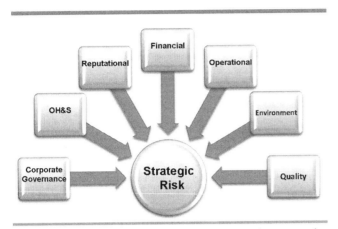

FIGURE 1. Interrelationship of strategic (business) risk and occupational safety and health risk (Global Solutions, Inc. 2011)

Leadership Commitment

Transparency and Outside Influences

Business risk management is influenced by the advent of government-led legislation and industry-led guidance, along with stakeholder social and environmental expectations, to improve financial transparency within domestic and multinational organizations. In the United States, the Sarbanes-Oxley Act of 2002 requires an organization to identify, control, and report financial risk to shareholders (U.S. Congress 2002). As a result, some U.S. multinational companies that are implementing a formal business risk assessment process are identifying and assessing health and safety risks that could significantly impact the financial integrity of the organization (Seabrook 2006b, Nash 2005). This business risk-assessment process expands all operations around the world, placing a company's brand integrity and business continuity implications at the forefront of risk identification and assessment (Nash 2005, ASSE 1996).

There are other Sarbanes-Oxley-type requirements in other countries around the world, such as the United Kingdom and Australia. In the United Kingdom, for example, the Turnbull Report, the Institute of Chartered Accountants' guidance document, *Internal Control: Guidance for Directors on the Combined Code*, was published in 1999. It proposed a risk-based approach, including internal control mechanisms, for the reporting of company risk to shareholders. The identification and reporting of significant health and safety risks is referenced in the guidance document. This guidance document applies to companies listed on the London Stock Exchange (Institute of Chartered Accountants 1999). There are similar requirements for trading on the Australian Stock Exchange (Australian Stock Exchange Corporate Governance Council 2003, Australian Securities and Investments Commission 2004).

In the financial markets, stakeholders are expecting a greater degree of socially responsible performance from the companies in which they invest. This includes social, environmental, and worker safety and health performance. Investors are looking to companies to be good corporate citizens, going beyond the legal requirements of safety, labor, or environmental regulations (European Agency for Safety and Health at Work 2004). Stakeholder expectation is beginning to emerge as a driver for greater health and safety risk transparency in multinational companies (Greenbiz 2005).

The number of companies publishing annual sustainability and corporate responsibility reports to stakeholders (which include health and safety performance) continues to grow. Sustainability—corporate social responsibility (CSR) and environmental, social, and governance awareness (as CSR is known in Europe) (Kaye 2010)—is beginning to impact business initiatives in European and U.S. companies. The number of companies identifying metrics and reporting their sustainability and corporate social responsibility performance has increased in the last 5 years. This increase is being influenced by several factors, including the investment community; customer requirements; "green" consumers; European Union, Chinese, and U.S. environmental regulations; supply-chain accountability; and the rise of regulatory and volunteer carbon emissions trading markets. Some companies see it as a competitive advantage (Seabrook 2010 and United Nations Environment Programme, KPMG Advisory N.V. 2010). Another incentive is the direct return on investment and operational efficiencies resulting from implementing environmental initiatives such as reduced packaging and transportation costs.

The Global Reporting Initiative™ (GRI) a European-based organization (Secretariat office: Amsterdam, The Netherlands) has developed a sustainability reporting framework, which is widely recognized, referenced, and used as a guidance framework for disclosing sustainability performance. The framework sets out performance indicators for disclosure as well as disclosure requirements to assure the quality and accuracy of the information disclosed in a performance report (Global Reporting Initiative™ 2006).

On the workplace safety and health side of sustainability performance reporting, the GRI framework lays out social performance indicators under "Labor Practices and Decent Work Aspect: Occupational Health and Safety" (Global Reporting Initiative™ 2006). Currently there are two core performance indicators associated with workplace safety and health sustainability. The first is injury, occupational diseases, fatalities, lost work days, and absenteeism rates. The second is risk control measures, including training and education, counseling, and prevention for serious diseases (Global Reporting Initiative™ 2006). While these are not optimal safety and health metrics in and of themselves, and occupational safety and health management systems would be better, these GRI metrics open the window of opportunity for safety professionals to engage in sustainability initiatives within their organizations. Some multinational companies such as Henkel (German-based) and Hewlett Packard (HP) (American-based) have gone beyond these core performance indicators to include areas such as risk management in the area of occupationally related motor-vehicle injuries and illnesses (Henkel 2009) and EHS management system performance (HP 2009).

For the safety professional, the environmental aspect of sustainability also has a direct impact on worker health and safety. Reducing or eliminating hazardous substances used in manufacturing processes reduces hazardous waste generated. Identifying and managing health and safety risks prior to implementing alternative energy solutions has a direct impact on a potential worker, as well as on the safety and health of the community, customer, and general consumer. The basic tools of risk and hazard identification, assessment, and control should be incorporated into the corporate (global) safety and health standards, as well as into the standard operating procedures at site locations around the world. For example, proactively assessing and managing the risks to workers involved in the routine maintenance of roof-mounted solar panels would reduce the potential for falls, burns, and muscular or skeletal injuries.

Another influence on transparency and identification and management of health and safety risk is the International Standards Organization (ISO) guideline *ISO 26000 Standard on Social Responsibility*. The joint secretariat for the proposed ISO 26000 standard is the Swedish Standards Institute and the Brazilian Association of Technical Standards (ISO 2010).

Integration

In order to effectively manage workplace safety and health risk, an organization's leadership must be committed to developing and implementing a system to manage this risk throughout their global organization (BSI 1999, Seabrook 1999a). By definition, an occupational safety and health management system (AIHA 2005, BSI 1999, ILO 2001) provides an internal control mechanism for workplace safety and health risk, a requirement under Section 404 of the Sarbanes-Oxley Act of 2002 (U.S. Congress 2002).

Workplace safety and health risk is one of many business risks an organization must manage. The author worked with a U.S.-based multinational company that had purchased a Swedish-based European multinational company, which manufactured and distributed components for the automotive industry throughout Europe and Scandinavia. The Swedish company's business model was based on a philosophy of one management system for all business risk. This meant that all functional areas of the business reported to one manager of management systems, who oversaw the planning, implementation, monitoring, control, and continual improvement processes for all of the company's business risk. Some examples of functional areas that this included were finance, human resources, research and development, marketing, sales, technical services, the environment, worker health and safety, quality, production, and distribution.

This demonstrates a truly integrated systems approach to business risk management. It also demonstrates that various safety and health risk management models work as long as the model and approach are aligned with the company's organizational structure, business model, and culture. The ANSI Z10, *Occupational Health and Safety Management* standard's guidance section suggests that health and safety risks associated with all functional areas of an organization are included in the risk management (identification and assessment) process (ANSI/AIHA 2005).

In order to successfully manage safety and health risk within a global company, the safety and health professional should understand the interconnectivity of local country cultural norms and their company's core business strategies, operational structure, and type of products and services, as well as the extent of worker safety and health risks. Understanding integration at this level will position the safety and health professional to successfully contribute to the financial, operational, and health and safety performance of their company.

Figure 2 depicts a global business model where worker safety and health risk-reduction strategies are integral to the business strategies, models, ethics, values, and organizational structure of an organization. The impact of local cultural norms is reviewed in the section, "Global Strategy–Local Implementation."

On a global basis, there is no typical U.S. business model for corporate workplace safety and health. Depending upon a company's organizational structure, the safety and health function may report to either the chief executive officer, chief financial officer, chief risk officer, global operations executive, legal department, or human resources. In many companies, the safety and health function is integrated with the environmental management function to form the corporate environment, health, and safety (EHS); health, safety, and environment (HSE); or safety, health, and environment (SHE) services. In recent years, some U.S. organizations have moved toward integration of worker-focused staff functions, such as employee well-being service centers.

While the name of the department may differ from company to company, it typically incorporates workplace safety and health, environmental issues, medical services, security, workers' compensation claim management, and sometimes overall risk management/insurance services. With the emphasis on security, antiterrorism, and emergency preparedness following the 2001 World Trade Center attacks in New York and the potential for global flu and other health-related pandemics, U.S. multinationals have focused on security and preparedness (emergencies resulting from natural disasters, fire, business contingencies, war, terrorism, and health issues) within the workplace safety and health function (Global Solutions, Inc. 2011).

The organizational structure, defined functionally with clearly delineated responsibility, accountability, and authority for workplace safety and health, is the key to effectively managing global workplace safety and health risks (Seabrook 1999). These areas of functionality, responsibility, accountability, and authority must be clearly communicated by the CEO throughout the business units and throughout the world. The communications should be written and include a corporatewide policy statement, backed by annual goals, objectives, and targets that are integrated into the global business plan and reported to stakeholders in the annual report. Global business units live and die by their projected goals, objectives, and targets. If managing global workplace safety and health risk is the CEO's expectation, and goals and targets are set, these risks will be controlled by the management team throughout every business unit and facility in the global organization (BSI 1999; ILO 2001; AIHA 2005; and Global Solutions, Inc. 2011).

FIGURE 2. Worker safety and health aligned with business strategies (Global Solutions, Inc. 2011)

Throughout the global organization, leadership commits to:

- Protecting the worker from work-related injuries, illnesses, disease, and death
- Integrating occupational safety and health management systems as part of the overall management of the organization
- Providing human, financial, and technological resources for safety and health management
- Complying with country-specific regulations
- Providing transparency and the communication of workplace safety and health risk information to all employees and stakeholders
- Including worker participation
- Improving safety and health performance

FIGURE 3. Leadership commitment (AIHA 2005, BSI 2008, Seabrook 2011)

Figure 3 outlines the areas of leadership commitment in managing global worker safety and health risks. At a minimum, these should be included in the organization's safety and health policy statement.

In order to clearly communicate the organization's worker safety and health expectations, the CEO and the leadership team must complete the following tasks (AIHA 2005; BSI 2007; ILO 2001; Global Solutions, Inc. 2011):

- Develop a written statement of policy, signed by the CEO, demonstrating leadership commitment to managing the organization's global safety and health risks (see Figure 3).
- Communicate expectations for managing workplace safety and health risks for both management and the workers throughout the global organization, using all communication media available (for example, the annual report to shareholders, the corporate intranet, and training and development documents).
- Commit resources to manage worker safety and health risks within the organization.
- At the corporate level, develop, communicate, implement, track, update, and report strategic goals, objectives, targets, and results of workplace safety and health risk-management processes.
- Integrate corporate goals, objectives, and targets into the organization's strategic business plan.
- Develop, communicate, implement, track, update, and report on annual goals, objectives, and targets at the global business unit level to manage workplace safety and health risks.
- Integrate each business unit's goals, objectives, and targets into its business plan.
- Develop performance-measured individual and group reward systems based on the results of goals, objectives, and targets, beginning with executive leadership at the corporate level, and moving down through the organization to the line management and workers within global business units.
- Integrate worker safety and health expectations into job descriptions and job procedures.
- Audit performance and have senior management review results at least annually.

Corporate safety and health policy globally is dependent upon the leadership's commitment and ability to translate the policy locally, tying worker safety and health performance to the business unit and local facility results (AIHA 2005; Global Solutions, Inc. 2011).

These business leaders must agree on worker safety and health goals, objectives, targets, and the organizations' values and key performance indicators in order for the worker safety and health process to move forward. Once this is accomplished, the corporate, business unit, regional, and facility-level safety and health managers, coordinators, and outside resources can develop business plans and allocate resources to carry out those plans.

Following the leadership commitment, organizational resource allocation is the next key factor in effective global worker safety and health management. In order to implement business plans, leadership throughout the organization must allocate resources to implement the worker safety and health aspects of the plans. This includes human, technological, informational, and financial resources. If resources are not committed at the highest levels, worker safety and health risks will not be managed (BSI 2007, AIHA 2005, ILO 2001).

Organizationally, having an effective corporate/business unit/regional/facility safety and health professional is a technical and managerial resource to leadership. The professional's expertise in assessing worker risk and providing fact-based information and solutions enables leadership to make informed decisions based on understanding all business risks.

Without senior leadership direction, worker safety and health is generally not integrated into the business plans for each business unit and facility. This is typically due to societal and business cultural norms in Asia, Latin America, Africa, and the Middle East, where worker safety and health strategies for an organization's facilities are not advanced at the grass roots level. This is also true in countries with a high level of poverty, where workers do not have a stable source of food, water, shelter, and personal safety available to them. In these countries, workers may willingly suffer poor working conditions because the alternative is extreme poverty (Global Solutions, Inc. 2011).

For example, in Hong Kong, the perception of risk acceptance in regard to construction scaffolding is different than in countries such as the United Kingdom, Australia, Germany, France, and Canada. In Hong Kong, the use of multistory bamboo scaffolding is a standard industry practice and meets regulatory standards under the Factories and Industrial Undertakings Ordinance and Construction Sites (Safety) Regulations (Hong Kong Labour Department 2001). Country-to-country, the perception of risk and risk acceptance may differ. It is hard to imagine contractors in the United States working on six-story bamboo scaffolds.

Therefore, without leadership at the business unit and facility level and organizational time and resource commitment, the safety and health professional will be wasting time, as well as the organization's resources, when attempting to conduct an audit at the organization's local facility without a proper introduction at the leadership level. For example, in India and Japan it is essential to be properly introduced to local management by company senior management before scheduling visits or audits.

The role of the corporate safety and health function is to assist leadership in developing and implementing a global safety and health management system, along with corporate standards, programs, internal reporting, and document control processes. Once developed, leadership must communicate these expectations throughout the global business units.

A GLOBAL SAFETY AND HEALTH RISK PROFILE/SCORECARD

A global safety and health risk profile (GRP) or a safety and health scorecard provides an organization's leadership with a current status of significant global safety and health issues. This tracking tool also assists leaders and safety professionals with managing workplace safety and health risk through an organization's strategic planning process. It enables the safety and health professional to actively engage senior management by integrating global workplace safety and health issues into a company's strategic risk-management framework. The GRP is comprised of an Executive Summary with supporting documentation, and identifies and assesses significant workplace safety and health risks throughout an organization's global operations. These significant worker safety and health risks pose a threat to human life and the economic and operational health of the company (Global Solutions, Inc. 2011).

The GRP is a dynamic document. It changes as the nature of the operations, risks, scope, and size of an organization change. Ideally, the GRP should be updated and communicated to senior leaders and management, including the Board of Directors, on a regular basis (Global Solutions, Inc. 2011).

The GRP incorporates a country-by-country, site-by-site analysis of significant safety and health risks. The company should use their global safety and health management process to identify and assess the management of significant safety and health risks throughout all global operations, including manufacturing, distribution, warehousing, offices, home offices, and fleets. The Executive Summary in the GRP provides a high-level economic and operational assessment of these significant safety and health risks and the potential economic impact they could have on the company.

For example, the improper handling or storage of hazardous materials at a multinational company's Chinese manufacturing operations could have a significant impact on the financial integrity of that company.

Without a hazardous materials management process in place, the storage of these materials presents a significant fire and explosion risk to adjacent manufacturing operations. If an explosion occurred in the storage area, it would cripple the main manufacturing and storage areas. Loss of the plant, people (workers and the public), or damage to the environment would affect public confidence and the company's reputation with the government and public officials. This could result in governmental intervention and work stoppage, fines, and litigation. It could also impact the timeframe to rebuild the facility due to permits and the ability to retain a contractor. If this manufacturing site is the sole supplier for the company's largest U.S. customer, loss of customer confidence could also impact current and future orders and revenue from that customer.

Another example of a significant worker safety and health risk that could impact the financial health of a business is asbestos liability. The scenario: during the initial safety and health audit of a recently acquired Mexican manufacturing company, the audit team discovers that along with the purchase of a manufacturing operation, the Mexican operations include a medium-sized asbestos removal contractor. This fact was not disclosed during the preacquisition due-diligence process. The asbestos health risk to past and present workers and the resulting financial implications for asbestos liability could be significant. This asbestos health risk is an example of the type of significant safety and health risk that should be assessed and monitored through the global safety and health risk profiling process.

From a regulatory standpoint, the United Kingdom's Financial Services Authority (a governmental oversight agency similar to the U.S. Securities and Exchange Commission) requires all financial services companies operating in the United Kingdom to develop a risk register. The risk register incorporates all risks that could negatively impact a company's financials and financial reporting, including environmental, health, and safety issues. United Kingdom financial services companies should incorporate the global safety and health risk profile into the risk register.

From an investor's perspective, nonfinancial reports provide an annual overview of a company's environmental, health, and safety risks. As early as 2005, the European Agency for Safety and Health at Work within the European Union stated, "corporate social responsibility (CSR) is becoming an increasingly important priority for companies of all sizes and types in 2005. Occupational safety and health is an essential component of CSR. . . ." (European Agency for Safety and Health at Work 2004, 2005). Although CSR is not regulatory-driven, stakeholder awareness of corporate social responsibility (sustainability) issues continues to grow as an economic driver for nonfinancial reporting to stakeholders. Companies such as BP, HP, Unilever, Ford Motor Company, Intel, Gap Inc., Johnson & Johnson, Henkel, General Mills, Royal Dutch/Shell, and Novo Nordisk recognize and report their environmental and social impacts in an annual or biannual nonfinancial report. As early as 2004, the United Nations Environment Program, along with the consulting firm, SustainAbility, who partnered with the credit rating agency Standard & Poor's, began to rank participating companies based on their nonfinancial environmental and social impact and performance. Standard & Poor's uses these reports to assess a company's risk profile (The Economist 2004). Therefore, there is an economic incentive for companies with good CSR/sustainaiblity performance in nonfinancial risk areas such as health and safety to disclose this information. Some companies see this as a competitive advantage and manage safety and heath risks accordingly. The Nike supply-chain management case study at the beginning of this chapter is an example of a significant safety and heath risk that would be identified and tracked by senior management as a significant business risk in a global safety and heath risk profile.

GLOBAL STRATEGY– LOCAL IMPLEMENTATION
Global Strategy

One of the greatest challenges a multinational organization may face is balancing global corporate goals, objectives, and targets with local, in-country infrastructure, facilities, knowledge, regulatory environment, culture, healthcare systems (government-run versus third-party insurance), capability, and competence.

Many multinational companies implement a corporate-driven global worker safety and health management system strategy throughout their facilities. This provides consistent set of standards, process, approach, and means for communication across the globe. This global safety and health management strategy is coupled with knowledge that there may be more than one way to manage a risk at a local facility, while being in compliance with corporate standards and guidance. The size of the organization does not matter, this strategy can be implemented within a global organization that has 5 or 115 sites around the world (Nash 2005; Global Solutions, Inc. 2011).

What does a centralized corporate management system look like? It will differ from one company to another, based upon the company's leadership commitment to workplace safety and health, corporate culture, size, type of operations, level of risk, and resource allocation. The management system may differ in organizational structure, terminology, methodology, and implementation. A management systems approach is discussed more thoroughly in the section, "Global Safety and Health Management Systems." From a strategic perspective, the occupational safety and health management system implementation should embrace the local culture and local regulatory environment.

Local Implementation: Cultural Impact

An example of thinking globally and implementing locally is seen in differing risk-control methodologies. For example, in operations in developing countries, where the cost of labor is inexpensive and equipment and facility costs are high, one is likely to see a very different way of doing work. In Indonesia, moving raw materials from storage to processing may mean employing several workers to do the manual material handling as opposed to automating the raw materials transfer or using manual handling equipment. Culturally, this is perceived to increase jobs, reduce expenses on the organization, and provide a safe methodology to do the work. Upon closer observation and questioning, it may be determined that the loads lifted are within the corporate safety and health lifting/manual handling standards and guidelines. Due to local culture and business practices, automating the raw materials transfer process may not be the best option for local leadership.

In another example, a multinational organization found that their operations in India used a state-of-the-art, continuous chlorine gas-monitoring system in their chlorine water treatment operations for the facility. Upon closer inspection and questioning, it was discovered that this monitoring equipment (an American model) had not been serviced or tested since its installation, and the technology and knowledge to test the system, calibrate the test equipment, and do preventive maintenance on the equipment was not available locally. The multinational organization replaced the American system with one that could be properly maintained and tested locally.

It is important to incorporate cultural elements into the management system.

A successful strategy for assessing the management of risk at local in-country facilities is to:

- Focus on outcomes: What is the risk? How is it managed? Is the risk adequately controlled?
- Allow for cultural, technological, legal, regulatory, healthcare, social system, and other differences in developing solutions to risk-control challenges, findings, and other implementation strategies. The best advice: Ask a lot of questions before mandating risk-control solutions for your global facilities (Global Solutions, Inc. 2011).

Local Implementation: Regulatory Environment

An important lesson in global safety and health management is to understand that the United States Occupational Safety and Health Administration (OSHA) does not have jurisdiction outside the United States; therefore, its regulations are also not recognized or, in some cases, known outside of the United States. Most developed countries in the world have their own set of regulatory requirements for worker safety and health. Just as the United States does not recognize the Korean Ministry of Labour as having jurisdiction

TABLE 1

Some Government Worker Safety and Health Regulatory Agencies

Country	Agency
Australia - New South Wales	Minister for Commerce - WorkCover New South Wales
Australia - Victoria	Victorian WorkCover Authority
Canada - Federal	Labour Program Human Resources and Skills Development
Canada - Ontario	Ministry of Labour - Occupational Health and Safety Branch
Mexico	Federal Secretary of Labor and Social Welfare (Secretaría del Trabajo y Previsión Social - STPS)
China	The Ministry of Labour and Social Security
India	Directorate General, Factory Advice Service and Labour Institutes, Ministry of Labour
Japan	Ministry of Health, Labour and Welfare - Industrial Safety and Health Department
Korea	Ministry of Labour
European Union	Directorate-General for Employment, Social Affairs and Equal Opportunities
United Kingdom	Health and Safety Executive
Germany	Bundesministerium für Wirtschaft und Arbeit
France	Ministère de l'emploi, du travail et de la Cohésion Sociale DRT

(*Source*: Global Solutions, Inc. 2011)

in the United States, so Korea does not recognize OSHA as having jurisdiction in Korea. Table 1 references some country-specific government regulatory agencies that oversee worker safety and health (Global Solutions, Inc. 2011).

The safety and health regulatory environment is different from country to country. In Germany, Singapore, the United Kingdom, and the European Union, some worker safety and health regulations are more stringent than in the United States. In other countries, such as Mexico, there are good regulations with limited resources for enforcement. There are other countries with limited or no regulations, such as Vietnam and Cambodia. In the European Union, for example, it is an organization's statutory duty to implement a system to manage worker safety and health risk. In Singapore, organizations are required to conduct self-audits and report findings to the regulatory agency. Safety committees are required in organizations with nineteen or more workers in Brazil (Secretaria de Segurança e Medicina do Trabalho 2004), and in the United Kingdom, eye testing is required of all significant display screen equipment users (HSE 1998).

In Canada, Part II (Occupational Health and Safety) of the Canada Labour Code imposes a legal duty on employers and "those who direct work" to take reasonable measures to protect worker and public safety. This legal duty has been codified in the Canadian Criminal Code, Bill C-45, which indicates that charges of criminal negligence can be made against an organization if death or bodily harm results due to "wanton or reckless" disregard of this legal duty. How this will impact individuals such as CEOs, managing directors, or executive leadership will be determined as cases go through the Canadian court system (Human Resources and Skills Development Canada 2005 and CCOHS 2010).

GLOBAL SAFETY AND HEALTH MANAGEMENT SYSTEMS

The challenge faced by all multinationals is to develop a management system or process that is valued and used by leadership to identify, communicate, implement, and monitor country-specific worker safety and health regulatory obligations, as well as global, internal corporate requirements for worker safety and health.

An organization must develop a system to manage its safety and health risks that is aligned with the business strategies of the organization. As described in the section, "Leadership Commitment," and illustrated in Figure 2, business strategies include business focus, culture and models, organizational structure, and philosophy. Within the U.S. pharmaceutical industry, for example, the U.S. Food and Drug Administration (FDA) approves all prescription drugs for the U.S. market, thus providing the key to the sustainability of a U.S. pharmaceutical company. Because of this, a pharmaceutical company's focus, culture, and organizational structure, from the research and development team to the operational aspects of the organization, are typically based on FDA approvals and standards of practice. Operationally, this translates into a corporate value and ethos of high standards for quality and hygienic production at their facilities. From the author's experience, those U.S. workplace safety and health professionals who understand this business culture and motivation, and who work to communicate

Worker Safety, Health, and Environmental Management System Standards

Australia/New Zealand (AS/NZS) 4801: 2001 *Occupational Health and Safety Management Systems-Specification with Guidance for Use*. Standards Australia.

American National Standards Institute (ANSI) *ASC Z10: 2005 Occupational Health and Safety Systems* (in development at the time of writing). American National Standards Institute.

Occupational Health and Safety Assessment (OHSAS), Series 18001: 2007. British Standards Institute.

International Labor Organization ILO-OSH 2001 *Guidelines on Occupational Safety and Health Management Systems*. International Labour Organization.

International Standards Organization (ISO) 14000: 2004 *Environmental Management Systems, Specification with Guidance for Use*. International Standards Organization.

Management of Health and Safety at Work Regulations, 1999 – Approved Code of Practice. United Kingdom implementation guidance on European Union Directive.

United States Occupational Safety and Health Administration. *Voluntary Protection Program*. OSHA.

FIGURE 4. Occupational safety, health, and environment workplace management systems standards, regulations, and models (*Source:* Global Solutions, Inc. 2011)

with leadership and integrate the management of worker safety and health risks with the management of product quality, are more successful.

Alternatively, the automotive industry is driven by lean manufacturing and efficiencies, and waste minimization is gained through product design and quality. Therefore, aligning and integrating worker safety and health management systems with operational, engineering, design, and quality systems proves more successful in this business environment.

It is not within the scope of this chapter to provide an in-depth review of Occupational Health and Safety Management Systems (OHSMS). However, Figure 4 is included to provide a quick reference for recognized, published OHSMS standards available to the safety and health professional. These OHSMS standards provide a template to develop and implement an OHSMS that is aligned with an organization's business strategy. Figure 5 provides the basic components of a graphic illustration of a global management system.

CONFORMITY ASSESSMENT: MANAGING CORPORATE AND EXTERNAL COMPLIANCE ASSURANCE

An organization's conformity-assessment and local regulatory compliance process applies to the corporate headquarters, all business units, and facilities throughout the world. The goal of conformity assessment is to ensure that the organization meets its external country-specific regulatory obligations and internal corporate requirements for managing worker safety and health throughout its global operations. The challenge is to develop an assessment process that will be used by leadership to assess and monitor conformance with predetermined local and corporate worker health and safety obligations and requirements.

There are three forms of conformity assessment that an organization should consider when developing its conformity assessment process:

- self-declaration, based upon the organization's execution, audit, and review of performance
- second party, where companies may audit their suppliers
- third-party certification by an accredited body (often done for ISO 9000 and 14000)

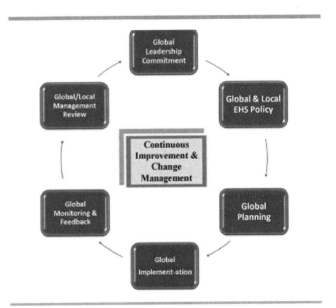

FIGURE 5. Basic components of a global safety and health management system (*Source:* AIHA 2005; BSI 2007; and Global Solutions, Inc. 2011)

In the experience of Stuart Wood, the head of the American Society of Safety Engineers' International Practice Specialty, companies that develop and implement good global safety and health standards, and apply these standards to all business units throughout the world, will be in compliance with approximately 90 percent of the local safety and health regulations under which a multinational operates around the world (Wood 2006). According to Wood, the most important strategy for an organization is "to assure the right safety and health people are in place" within a company's business structure, and throughout the company's global organizational structure. He emphasized that the safety and health organizational structure should be aligned with the company's business structure to be successful (Wood 2006). For example, does the business operate regionally using an American, Asian Pacific, European, African, or Middle Eastern regional model, or is it organized by other geographical parameters or by type of business? An example of safety and health organizational structure by type of business is seen in the petrochemical industry, where there may be separate safety and health management structures deployed for global exploration, production, drilling, and manufacturing business units.

With the right people in place, a good regional or countrywide safety and health management team will assess safety and health regulatory compliance by identifying gaps in the corporate global safety and health standards, and ensuring that the local business unit is in compliance with both the corporate global safety and health standards and local regulatory safety and health requirements. Having corporate safety and health standards essentially levels the playing field for safety and health expectations around the world (Wood 2006).

Strategies for Developing a Conformity Assessment Process

Conformity assessment strategies focus on developing and implementing a global corporate conformity assessment process that involves understanding local safety and health regulations and reporting requirements while allowing for how a culture may impact the implementation of corporate safety and health management systems and programs.

Trained corporate assessment teams conduct the review, and results are reported to the leadership of the organization and business units. For larger organizations, this is often part of the overall governance process of an organization. Assessment teams will benefit from an educational program on working internationally, specifically on how a country's local culture can impact how business is done in facilities outside their home country. There are many cross-cultural resources available. An organization's human resources department usually has access to this information.

The assessment results are a scorecard on how well a facility or business unit is doing relative to managing their worker safety and health risks. There is no one *best* assessment tool. Some companies use a numerical scorecard system to identify the level of compliance with corporate requirements, while others us a color-coding system. Both have become accepted practices over time. What is essential is that the system used has value to company leadership in managing their global safety and health risks. Figure 6 is an example of a color-coded scorecard system. Many safety and health consultants and multinational companies use these universal colors to indicate action in their conformity assessment process. The premise is analogous to a stop light. Green depicts safety and health risks are controlled and that no immediate action is needed. Yellow means caution, that safety and health risks may not be completely controlled,

RED
Significant (imminent danger, loss of life, property or damage to the environment) Safety and Health (S&H) risks exist and are not controlled; no S&H systems are in place.

YELLOW
Significant S&H risks exist but are controlled; some non significant, uncontrolled S&H risks exist; most S&H systems are in place.

GREEN
No significant S&H risks exist; all S&H systems are in place.

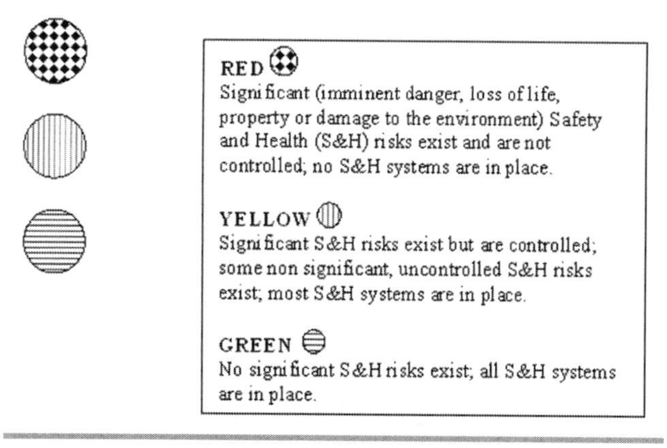

FIGURE 6. Sample score card (*Source:* Global Solutions, Inc. 2011)

but they are not so significant as to require immediate action. Finally, red means that safety and health risks are significant and that action is required immediately (Global Solutions, Inc. 2011).

Assessments often focus on the evaluation of the company's occupational safety and health standards and processes in addition to an audit of the workplace. Assessment frequency will be different for each facility, based on its size, nature of operations, number of workers, and the level of risk to workers, the public, the environment, and the business.

The assessment tool itself determines whether an organization is meeting its worker safety and health regulatory obligations and internal requirements. The following provides examples of these requirements.

Country-Specific Worker Safety and Health Regulations

Country-specific worker safety and health regulatory requirements may be stricter than the corporate standards in areas such as electrical safety, forklift driver certification, safety committees, medical monitoring, and hazard communication. In addition, regulatory reporting requirements may apply, such as worker safety and health-related incidents and accidents, self-audit results, use of specified hazardous materials, dangerous occurrences, spills or other releases of hazardous materials. An example of a country-specific regulatory requirement is the Canadian Workplace Hazardous Materials Information System (WHMIS). This legislation applies to all manufacturers, distributors, and users of specified hazardous materials, and incorporates labeling, material safety data sheets, and training requirements for workers using or being exposed to hazardous materials identified in the WHMIS regulations (Health Canada 2010). In the United Kingdom, the Reporting of Injuries, Diseases and Dangerous Occurrences Regulations, 1995 (RIDDOR) require dangerous occurrences to be reported to the regulatory authority when they occur. Dangerous occurrences include the malfunction of a breathing apparatus, collapse of scaffolding or an overhead crane, or turning over a forklift truck (HSE 2008). In Korea, it is an employer's duty under the Industrial Health and Safety Act (amended September 22, 2006) to assign a facility safety and health management officer to oversee implementation of the worker safety and health management system (OSHA 2009).

CORPORATE SAFETY AND HEALTH MANAGEMENT SYSTEMS

Corporate safety and health management system requirements include, but may not be limited to, management system components such as safety policies, communications systems, safety and health performance requirements in all worker job descriptions and annual performance goals, a risk assessment process, annual management reviews, self-assessment processes, document control processes, and change management.

Corporate Safety and Health Standards and Country-Specific Safety and Health Regulations

Corporate worker safety and health standards include facility and equipment self-inspections, lockout-tagout processes, forklift driver certification and licensing, incident investigation, safety committees, confined-space-entry permitting, medical monitoring, preventive maintenance action item tracking, safety and health training and retraining, and contractor safety and health management control and monitoring systems. These corporate safety and health standards may be in compliance with country regulations. In many cases, implementing the corporate safety and health standards will ensure "90% compliance with most country specific regulations" (Wood 2006). If a country does not have a developed worker safety and health regulatory environment, the corporate safety and health standards provide the framework for governing safety and health practice for the company in those countries. If a country does have a mature worker safety and health regulatory environment, where these regulations are stricter than the corporate safety and health standards, local personnel must be responsible and accountable for identifying and implementing these additional measures to assure compliance (Avon 2011).

According to L'Oreal's 2008 sustainability report, it is a consumer products manufacturer with 64,600 employees operating in 130 countries around the world. The company has implemented a corporate safety and health management system. All U.S. sites are part of the OSHA VPP program; other sites (with the exception of two) are Occupational Health and Safety Assessment Series (OHSAS) 18001 accredited. L'Oreal also piloted a joint safety management program with senior managers instead of graduate school for business in France (L'Oreal SA 2008).

Resources

Whether a large, medium, or small organization, local external safety and health resources can provide the assessment team with important in-country expertise on the implementation expectations of local regulatory agencies. The best source for good safety and health resources is from other global organizations, as well as through industry-specific associations, professional societies, and personal networking. The Web is another good means for identifying resources; however, a word of caution—always get good references and follow up on them. For example, when a multinational, service-based industry, operating 300 facilities in 23 countries around the world, needed to find a vendor to do thermal imaging for their electrical panels at their India facility. The global safety and health director eventually found an India-based vendor from the Internet, and the thermal imaging task was successfully completed. The key to success was having the local facility management vet the potential vendor. They knew the local market, could speak the local language, and reviewed and contacted the vendor's references to assure its credibility.

Appendix A highlights a number of online resources that may be useful in identifying in-country worker safety and health regulations and requirements.

Identify Key Global Safety and Health Performance Indicators

Another challenge that a multinational organization experiences is developing meaningful safety and health performance criteria that are applicable to all of their global business units. These indicators must be recognized by leadership and reflect an accurate measurement of a facility or a business unit's worker safety and health performance.

When managing safety and health globally, it is important to use both proactive (leading) and reactive (lagging) performance indicators. Using country-specific accident/incident statistics (a lagging indicator) will not always provide an organization with accurate, comparable performance data. Therefore, it is important to use proactive performance indicators as well. Some examples of performance indicators used in business include the percentage of audit findings closed based on a site action plan, the number of workers trained in confined space entry against the target in the annual worker training plan, and the percentage of risk assessments completed against the targets set in the annual plan. It is important to normalize raw numbers. For example, the number of injury, illness, or driving incidents should be normalized against the number of employee hours worked, the number of finished products, or the total number of miles traveled. This provides consistency and value to statistical data.

Statistical data variables and calculations for injuries and illnesses vary from country to country. For U.S. domestic operations, OSHA recordable incident definitions apply as well as OSHA's lost-workday injury calculations. Once outside the United States, lost-time case rate definitions and calculations may vary from country to country. In most cases, U.S. facilities would typically have a higher lost-time case rate than, for example, their Japanese and EU counterparts, due to a more conservative definition of a lost-time case.

Specifically, discrepancies in lost-time case rates (LTCRs) country-to-country are due to differences in how they are defined, calculated, and reported to the local regulatory agencies. In an organization's Japanese facility, for example, the LTCR could be artificially lower than that in their similar UK, EU, or U.S. facilities. Figure 7 illustrates the number of days a worker must be away from work (due to a workplace injury) for the injury to be classified as a lost-time case. In Japan, a lost-time case is reported to the regulatory agency after a worker has been away from work seven days following a workplace injury or illness. In the EU, it is after three days, and in the United States, it is after one full day away from work. Therefore, a company using a consistent LTCR calculation globally (for example, the number of lost-time cases per 200,000 hours worked) will not find an accurate comparison of performance because each of the above countries defines a lost-time case differently.

Another issue that can influence the LTCR is the healthcare system practiced in a country. In countries where a government-run socialized health system is in place (e.g., Spain, Italy, and France), it is more difficult, or not legally possible, to implement a return-to-work program. Again, an accurate comparison of performance site-to-site or country-to-country is impacted.

To overcome discrepancies in incident reporting, and to develop meaningful data, multinational organizations require their facilities to report two sets of accident data and statistics. One set is developed to comply with the local regulatory agency's reporting requirements, and the other for corporate reporting to track internal safety and health performance for global comparison.

Accurate local incident data and reporting can impact government oversight and action as well. In the United Kingdom, the Reporting of Injuries, Diseases and Dangerous Occurrences Regulations, 1995 require specific incidents to be reported to the Health and Safety Executive (the United Kingdom Health and Safety regulatory body) within 24 hours following the incident. The type and nature of the incident determines whether a visit from the Health and Safety Executive ensues (HSE 1998).

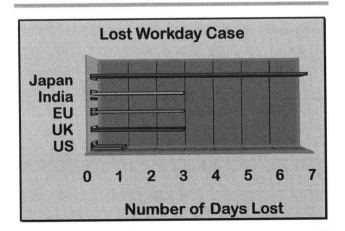

FIGURE 7. Defining lost-workday cases by country
(*Source:* Global Solutions, Inc. 2011)

Examples of Safety and Health Performance Indicators

- Improvement of corporate worker safety and health audit/assessment scores (quarter to quarter/ year to year)
- Percentage of corporate audit findings closed, by quarter
- Percentage of self-inspections completed on time, by a competent inspector
- Percentage of regulatory inspection findings closed one month following the date of inspection
- Percentage of retraining conducted on schedule
- Percentage of incidents investigated
- Percentage of risk assessments completed
- Percentage of risk assessment action items completed
- Percentage of safety and health goals, objectives, targets, and milestones completed

FIGURE 8. Safety and health performance indicators

To increase the likelihood of receiving accurate, meaningful incident rate performance data, an organization should develop and communicate written corporate reporting requirements, including:

- defined terminology: reportable (if used), lost time, incident
- defined timeframes for a lost-time case (e.g., worker away from work following an injury or illness after one full day, two full days, and so on)
- the accident rate calculation method (for example, based on 200,000 or 1,000,000 man-hours worked)
- written reporting procedures (local and corporate)
- developing and disseminating incident reporting forms
- other requirements based on individual organizational needs

Figure 7 illustrates lost-workday cases by country. Performance metrics define the expectations of an organization and allow the organization to measure its performance against set worker safety and health goals, objectives, and targets at the facility, regional, business unit, and corporate levels. Figure 8 provides some examples of proactive measurement criteria. They are proactive because they provide leadership with a measure of worker safety and health performance before an incident occurs. The intent of proactive measurement criteria is to identify management system deficiencies, provide corrective measures, and prevent an incident from occurring. Proactive indicators should include quantitative as well as qualitative measurement of performance.

Safety and Health Resources: Competency and Expertise

Sourcing competent local safety and health resources is essential. Local safety and health resources (ASSE 1996; Global Solutions, Inc. 2011):

- provide local regulatory and enforcement expertise
- understand the company's organizational structure
- know how to effectively implement systems, policies, and programs locally
- can assist corporate safety and health in translating local safety and health issues (such as why audit findings are not being closed)
- mitigate local language barriers
- understand local cultural norms, which influence effective implementation of safety and health systems, policies, and programs

Identifying and deploying these resources is crucial to effectively implementing a global safety and health process.

Needs Assessment

A needs assessment assists in identifying safety and health resource requirements for the facility or global region. The needs assessment should be based on an evaluation of facility locations, both regionally and locally, individual facility size, financial resources, nature of operations, and level of safety and health risk for the facility.

The following provides examples of how some multinational companies have deployed safety and health resources, including internal dedicated employees, internal shared (part-time) employees, and external consultants (Global Solutions, Inc. 2011):

- Regional—full-time/facility-shared resource: three light assembly facilities located in China, where the level of worker safety and health risk is low to medium at each facility. The organization employs a regional safety and health manager who oversees safety and health activities using internal, shared, local safety and health employees at each facility within the region.
- Facility—full-time resource: a 300-employee, high-risk ammonia processing facility. The organization employs two full-time safety and health employees.
- Facility-shared resource/external resources: an office with 50 employees. The organization employs one internal shared safety and health coordinator and uses external resources to manage specific safety and health issues, such as ergonomics, home office workers, electrical safety, fire, life safety, and fleet. The safety and health coordinator is a shared resource with human resources.

Regardless of how an organization assesses staffing needs, all safety and health resources must have the competency to fulfill their safety and health role within the organization.

Competence

A needs assessment identifies safety and health competency levels required at the facility and regional levels. Competence is broken down into the specific skills and knowledge required to do the job. Skills and knowledge include risk identification and assessment; accident investigation; integrated management systems; and specific risk-control processes such as confined space entry, contractor safety, ionizing radiation, lockout/tagout programs, forklift truck safety, asbestos abatement, hazardous chemicals, hazard communication, and transport (ASSE 1996).

Competence can also be demonstrated through third-party certification or registration, examination, university or college matriculation, on-the-job training, and/or experience.

Tips for Identifying Competent Safety and Health Resources (Internal and External)

- Certification bodies have a registry of verified (certified/registered) third-party safety and health professionals, by country.
- Contact your industry trade groups for recommendations.
- Consult international safety and health professional networks, such as the American Society of Safety Engineers (ASSE) International Practice Specialty or the Institution of Occupational Safety and Health (IOSH) International Specialist Group.
- Professional safety and health organizations such as ASSE, IOSH, the Singapore Institution of Safety Officers, the Safety Institute of Australia, the Canadian Society of Safety Engineers, and the International Network of Safety Practitioner Organizations.
- Contact other multinational companies working in the region or country where resources are needed.

FIGURE 9. Finding global health and safety resources (*Source:* Global Solutions, Inc. 2011)

Recognized third-party competence designations include (Global Solutions, Inc. 2011):

- Certified Safety Professional (CSP) - American Board of Certified Safety Professionals
- Certified Industrial Hygienist (CIH) – American Board of Industrial Hygiene
- Charter Member of the Institution of Occupational Safety and Health (previously Registered Safety Practitioner, RSP) (UK)
- Canadian Registered Safety Professional (CRSP) - Board of Canadian Registered Safety Professionals
- Registered Safety Professional [RSP (Aust)] - Safety Institute of Australia

Some government legislation dictates competency requirements for operations and supervisory- and managerial-level staff within an organization. This is true for Singapore, where the Ministry of Manpower, in conjunction with the Singapore Workforce Development Agency, published formal occupational health and safety competency standards (Singapore Ministry of Manpower and Workforce Development Agency 2006).

Figure 9 provides some common tips used by global safety and health professionals to identify competent internal and external safety and health resources.

Once safety and health resources are identified, it is imperative to review their resumes/curriculum vitae and to check references thoroughly. Outside the United States, references are generally more readily provided. Language capability is also important. The resource should be fluent in English as well as the local dialect of the facility's management and employees. Dialects can pose a language barrier. One final area to consider is the nationality and religious compatibility of the safety and health resource and the country in which the individual will work. This can be a significant barrier in accomplishing the safety and health objectives intended by the safety and health resource when hired. To mitigate this potential issue, assure that local national management interviews the potential candidate before hiring, and, as a corporate safety and health leader, listen and work with the local facility if you are told the candidate is not suitable.

GLOBAL COMMUNICATION: THE KEY TO EFFECTIVE IMPLEMENTATION

Communicating corporate values and setting expectations for workplace safety and health are all part of good global safety and health management. Operationally, this includes communicating processes, procedures, executive-level safety and health tracking reports (e.g., the global profile), best practices, and loss-trending and prevention information.

Today, global communication is 24/7, 365 days a year via new and evolving technology capabilities. Use of Web-based platforms for information sharing, video conferencing, and access to sophisticated incident and risk-trending software promotes teamwork and a focused approach to data and people-driven management of safety and health risk. It is worthwhile to mention there are instances where face-to-face meetings are the most effective communication tool. Some examples may include one-on-one transfer of critical information or skills—a relationship needs to be developed before process/procedure/system implementation can occur—or when the senior leader requests your presence.

The safety professional should look to identify the most effective communication tools available within the organization, choosing the best tool based on the nature and purpose of the communication.

INTERNATIONAL TRAVEL: A PERSONAL RISK ASSESSMENT

This chapter would not be complete without a discussion on global business travel risks. Key factors employees should assess for work-related international travel risks include the destination of travel, political climate, work activities, climate and weather conditions, access to medical services, and length of stay in the destination country. The following outlines considerations for all company employees who travel outside their home country—including the safety professional—to ensure safe arrival and return from an international destination (Institution of Occupational Safety and Health (IOSH) 2007; Global Solutions, Inc. 2011). Table 2 presents an example of an international pretravel checklist. Employee travel risk should be part of the overall company safety and health risk assessment process. In many companies this is managed by human resources and the corporate or local travel departments. Employee training on global travel risks should be incorporated into the onboarding process, and guidance should be provided in the employee handbook.

Personal Security/Terrorism

Know where and when you will be traveling and the time of year. To assess your personal security and determine whether you should travel to a country, do your homework. Are you traveling to a destination where you may be impacted by civil unrest, be subject to a natural disaster (such as a tsunami, hurricane, or cyclone), or become a target of terrorism due to your nationality? The final decision to travel internationally rests on your shoulders. The deciding questions to ask are whether the international travel risk is too great and whether there is a high probability of injury, illness, or death? If the answers are yes, do not travel.

Personal security should be considered when dressing. Leave expensive jewelry, watches, and so on at home, as it makes you a greater target for theft.

TABLE 2

International Pretravel Checklist

1.	Do you know the countries to which you are traveling? Does the time of year coincide with that of natural disasters (tsunami, cyclones, hurricanes, tornadoes)?	YES	NO
2.	Do you have a valid passport?	YES	NO
3.	If a visa is required to enter the destination country, do you have a valid visa?	YES	NO
4.	Do you keep copies of your passport with visas with you while traveling? At a secondary location (for example, at home and office)?	YES	NO
5.	Has your physician assessed your medical capability for international travel?	YES	NO
6.	Are your immunizations up to date? Are they adequate for the destination country to which you are traveling?	YES	NO
7.	Do you have Certificates of Immunization where required for destination countries?	YES	NO
8.	Do you have a copy of emergency contact numbers for your destination country and country of origin?	YES	NO
9.	Do you know how to access medical facilities in your destination country?	YES	NO
10.	Does your company subscribe to an emergency service? Do you have an ID card for the service?	YES	NO
11.	Has ground transportation been arranged prior to travel?	YES	NO
12.	If you require prescription drugs, do you have enough for the duration of your trip? Are they in their original containers?	YES	NO
13.	If recommended by your doctor or company travel service, do you have a medical kit with antibiotics, clean syringes, and needles, if required?	YES	NO
14.	Are your clothes appropriate for the anticipated climate (for example, cold or tropical), including sunglasses with ultraviolet protection from the sun?	YES	NO
15.	Do you have a valid drivers license to operate a vehicle in the country where you are traveling?	YES	NO

(*Source*: Institution of Occupational Safety and Health 2008; Global Solutions, Inc. 2011)

There are many government and private consultancies that provide timely security information on go and no-go zones around the world. Your own company may have access to or contractual arrangements with some of these consultancies. See the list of resources at the end of this section.

Emergency Contacts

Keep an itinerary with contact numbers for each country and facility you are visiting on your trip. Also keep emergency contact information for your corporate headquarters, and family members who should be contacted in the event of a medical or other emergency. Make sure you have 24-hour emergency contact information, as the time zone differences may mean you need to contact corporate headquarters outside of business hours. Also, keep a list with contact information for the embassies or consulates in the countries and areas in which you are traveling. The U.S. State Department Web site has contact information for U.S. embassies around the world. See the Web site Resources section at the end of this chapter for more information on the U.S. State Department.

Transportation

Make arrangements for ground transportation prior to arrival in the destination country, and have local company staff arrange for your transportation from the airport to your hotel and the office. If your company permits employees to drive in the destination country, the following considerations should be incorporated into the employee travel policy:

- Make sure you hold a valid driver's license to drive in the country.
- Know the local automobile insurance requirements.
- Know your legal rights if in an accident.
- Know who to contact within the company in an emergency.
- Consider how to deal with communication and language barriers.
- Understand the meaning of road signs.
- Understand local traffic regulations.
- Preplan your route, avoiding dangerous travel areas.
- Have local currency ready for tolls.

A local driver vetted by your company knows the language, roads, customs, tolls, or gratuities needed to safely get you to where you need to go. Unless you are fluent in the language, customs, and laws, leave the driving to a local national. A high-risk situation once occurred when a safety professional from a large multinational consumer goods manufacturing organization was traveling to his facility in Central Mexico. The security risk was considered high, and the company provided him transportation by armed guards. While traveling from the company facility to the airport, their security vehicle was stopped by armed bandits. In this case, the bandits only wanted money, so the driver paid the "toll," and they were allowed to leave unharmed.

Automatic Teller Machines (ATMs)

The American Bankers Association (ABA) recommends that ATM users ensure they are in a well-lighted, secure area before using an ATM. This applies no matter where you are in the world (from New York City, U.S., or Bangkok, Thailand). In the past, kidnapping was a risk limited to a company's executives. In Latin America, more junior-level individuals have been detained and forced to withdraw the daily maximum amount from their ATM account until it reached a zero balance. Always be aware of your surroundings and travel in groups (ABA 2006).

An Exit Strategy

If you are going to places where there may be civil unrest, a political crisis, terrorism, other security warnings, or the threat of a natural disaster, you should contact your local embassy upon entering your destination country and provide them with local contact information. The embassy will keep you up to date in the event of a natural disaster or security issue that requires evacuation or additional security measures.

Do you have an exit strategy from your destination country? Always know the best means to leave the country in the event of an emergency. What are your air, land, and sea options? How will you get to the airport, port of call, or car rental center?

The United States government provides a pretrip registration service for U.S. citizens who would like to register with the nearest U.S. embassy or consulate in the destination country. This enables the government to contact the traveler in the event of an emergency that may significantly impact their life or travel.

The State Department Web site provides tips for travel outside the United States, document requirements, and passport and visa information to U.S. citizens (www.travel.state.gov).

Medical Concerns

Take your health seriously prior to and during international travel. Many travelers have died or become seriously ill from traveling internationally due to inadequate immunizations, lack of local medical knowledge, facilities, prescription drugs, clean hypodermic needles, medical evacuation availability, and procedures carried out in the event of a medical emergency. In addition, poor food hygiene can be the cause of dehydration and serious illness.

International travelers should have a physical examination and make sure that immunizations are up to date prior to commencing international travel. If you have a medical condition, have your physician assess travel requirements relative to the medical condition. Keep a letter with you at all times describing your medical condition and signed by your doctor.

Have your corporate medical team or a local international travel medical center do an immunization profile based on your anticipated travel. This will compare your current immunization status with required immunizations for prevalent diseases in the countries to which you are traveling. Common diseases include typhoid, hepatitis A and B, tetanus, malaria, diphtheria, yellow fever, and meningococcal meningitis.

Some countries require certification of immunization prior to gaining entry. Know what countries require these, and keep them with your travel documents.

Prior to departing from your home country, find out what medical services will be available to you at your destination, and what payment method your company has in place for medical services rendered? Most companies arrange this for you, but it is always best to confirm this information in writing before departure.

The U.S. Centers for Disease Control or your own company travel department will provide immunization requirements for pretravel. Make sure you have enough time to take all the courses of immunization before your trip. In some cases, the immunization process can take three months.

Work Travel

Passports should be up to date. Some countries now require the traveler to have held the passport six months in advance of travel. Know where you are traveling, and make sure your travel itinerary allows transit from one country to another. For example, if one has recently visited Israel, some Arab countries may not grant entry; the same may hold true in the reverse situation.

Plan your travel well in advance and know where visas are required for entry to a country. Visa requirements to enter a country are based on the country from which the traveler (traveler's passport) originates. Your company's corporate travel department or external travel resource should be able to secure a visa for travel well ahead of the date of planned travel. This will allow for any unplanned delays in the visa procurement process, such as the embassy requiring additional documents prior to issuing a visa. If you do not possess the required visa for entry into a country, you may be denied entry onto the plane or worse, denied entry to the country to which you have just traveled 22 hours by airplane.

Travel Resources

- Institution of Occupational Safety and Health Safety: *Safety in the Global Village. International travel safety assessment and guidance.* (www.iosh.co.uk/information_and_resources/guidance_and_tools.aspx)
- United States Department of State (www.travel.state.gov)
- American Society of Safety Engineers' global seminars (www.asse.org)
- The Centers for Disease Control and Prevention (CDC) Web site (www.cdc.gov) provides up-to-date medical advice for travelers around the globe. Includes information on countries with epidemics.

CONCLUSION

Managing worker safety and health risks is good business. In some cases it is a license to do business in a country; in others, it is the survival of a company when a safety and health incident occurs. Managing worker safety and health risks is a critical part of the overall strategic planning process companies engage in when running a global organization. The effective management of worker safety and health as a business risk has a positive impact on profit, market share, brand, and manufacturing and distribution within multinational organizations.

REFERENCES

Adidas Group. 2005. *Connected by Football: Social and Environmental Report 2005.* Germany: Adidas Group.

Allars, Kevin. "BP Texas City Incident: Baker Review." *Health and Safety Executive.* December 2007 (retrieved May 29, 2010). www.hse.gov.uk/leadership/baker report.pdf

American Bankers Association (ABA). 2006. *ATM Safety Tips* (retrieved August 9, 2006). www.aba.com/ Consumer Connection/CNC_contips_atm.html

American National Standards Institute and American Industrial Hygiene Association (ANSI/AIHA). 2005. ANSI/AIHA Z10-2005, *Occupational Health and Safety Management Systems.* Fairfax, Virginia: AIHA.

American Society of Safety Engineers (ASSE). 1996. *Scope and Functions of a Safety Professional* (retrieved June 3, 2010). www.asse.org/hscopa.html

Australian Securities and Investments Commission. 2004. *The Corporate Law Economic Reform Program* (Audit Reform and Corporate Disclosure) Act 2004 (CLERP 9). Australian Securities and Investments Commission.

Australian Stock Exchange Corporate Governance Council. 2003. *Principles of Good Corporate Governance and Best Practice Recommendations* (retrieved March 2003). www.asx.com.au/supervision/governance/index.html

Avon. 2011. *Global Safety Strategy,* #6 (retrieved June 30, 2011). www.avoncompany.com/corporatecitizenship/corporateresponsibility/whatwecareabout/workplace-safetyandhealth/safetystrategy.html

British Petroleum p.l.c. 2005a. *A Message from BP's Group Chief Executive* (retrieved May 31, 2005). www.bp.com

_____. 2005b. *Health and Safety* (retrieved January 25, 2005). www.bp.com/liveassets/bp_internet/globalbp/STAGING/global_assets/downloads/E/ES_new2005_safety.pdf

_____. 2005c. *Making More Energy: Sustainability Report 2005.* United Kingdom: BP p.l.c..

_____. 2005d. Press Release: *BP Products North America Accepts Responsibility for Texas City Explosion* (retrieved May 31, 2005). www.bp.com/genericarticle.do?categoryId=2012968&contentId=7006066

British Standards Institute (BSI). 2007. *Occupational Health and Safety Assessment Series 18001-2007: Occupational Health and Safety Management Systems-Specification.* London, England: British Standards Institute.

_____. 2008. Occupational Health and Safety Assessment Series 18002: 2008, *Occupational health and safety management systems – Guidelines for the implementation of OHSAS 18001:2007.* London, England: British Standards Institute.

Brown, Garett. 2004. "Vulnerable Workers in the Global Economy." *Occupational Hazards* (April) pp. 29–30.

Browning, Jackson B. 1993. "Union Carbide: Disaster at Bhopal." Gottschalk, Jack A., ed. *Crisis Response: Inside Stories on Managing Under Siege.* Detroit, Michigan: Visible Ink Press.

Burgess, Kate, and Rebecca Bream. 2006. *Investors Seek BP Safety Assurances* (retrieved September 14, 2006). www.ft.com/cms/s/0/ee78d5ce-4455-11db-8965-0000779e2340.html

Canadian Centre for Occupational Health and Safety (CCOHS). 2010. "Bill C45-Overview" (retrieved June 1, 2010). www.ccohs.ca/oshanswers/legisl/bill c45.html

China Association of Development Zones. 2002. *National Export Processing Zone* (retrieved January 14, 2005). www.cadz.org.cn/en/kfq/jj.asp?name=EPZ

Corporate Responsibility Magazine. 2010. "Corporate Responsibility Magazine: 100 Best Corporate Citizens for 2010" (retrieved March 3, 2010). www.thecro.com/files/CR100Best3.pdf

European Agency for Safety and Health at Work. 2004. *Issue 54: Corporate Social Responsibility and Occupational Safety and Health* (retrieved May 6, 2006). www.osha.europa.eu/publications/factsheets/54/index.htm/view?searchterm=issue%2054

European Agency on Health and Safety at Work. 2005. *Introduction: Corporate Social Responsibility and Safety and Health at Work: Volkswagen, Automobiles (Germany)* (retrieved August 4, 2006). www.osha.eu.int/publications/reports/210/index_24.html

Friedman, Thomas L. 2005. *The World Is Flat.* New York: Farrar, Straus and Giroux.

Global Reporting Initiative™. 2006. "G3 Sustainability Reporting Guidelines" (retrieved March 3, 2010). www.globalreporting.org/ReportingFramework/G3 Guidelines

Global Solutions, Inc. 2011. *Global Environmental, Health and Safety Management Seminar Workbook.* Mendham, New Jersey: Global Solutions, Inc.

Gordon, Margery. 2001. *Advantage Reebok* (retrieved October 10. 2006). www.calbaptist.edu/dskubik/rebok_nike.html

Greenbiz. 2005. *UN Global Compact Participants Report Progress So Far* (retrieved July 19, 2005). www.globalpolicy.org/reform/business/2005/0719gcreport.html

Health and Safety Executive (HSE). 2008. *A Guide to the Reporting of Injuries, Diseases and Dangerous Occurrences Regulations (RIDDOR) 1995, L73a.* Sudbury, England: HSE Books.

Health Canada. 2010. *Workplace Hazardous Materials Information System* (retrieved May 31 2010). www.hc-sc.gc.ca/ewh-semt/occup-travail/whmis-simdut/index-eng.php

Hewlett Packard (HP). 2006. *2006 Global Citizenship Report*. California, USA: HP.

———. 2008. "Working Safety, Staying Healthy." *2008 HP Corporate Social Responsibility Report* (retrieved March 3, 2010). www.hp.com/hpinfo/globalcitizenship/gcreport/employees/health.html

———. 2009. *HP Global Citizenship Report 2009*. California, USA: HP (retrieved June 1, 2010). www.hp.com/hpinfo/globalcitizenship/index.html

Hong Kong Labour Department. 2001a. *Construction Sites (Safety) Regulations*. Hong Kong: Hong Kong Labour Department.

———. 2001b. *Factories and Industrial Undertakings Ordinance*. Hong Kong: Hong Kong Labour Department.

Howell, Karen. 2004. "Working Abroad: Foreign Offices." *Safety and Health Practitioner* (August), pp. 27–30.

Human Resources and Skills Development Canada. 2005. *Occupational Health and Safety in Canada – Legislative Changes from September 1, 2003 to August 12, 2004* (retrieved May 28, 2005). www.hrsdc.gc.ca/en/lp/spila/clli/dllc/15_2003_2004.shtml#iii_a

Institute of Chartered Accountants. 1999. *Internal Control—Guidance for Directors on the Combined Code for Corporate Governance*. England: Institute of Chartered Accountants.

Institution of Occupational Safety and Health (IOSH). 2008. *Safety in the Global Village* (retrieved May 20, 2010). www.iosh.co.uk/information_and_resources/guidance_and_tools.aspx

International Labour Office (ILO). 2001. *ILO-OSH 2001 Guidelines on Occupational Safety and Health Management Systems*. Geneva, Switzerland: ILO.

International Standards Organization (ISO). 2006. *Guidance on Social Responsibility* (retrieved December 10, 2006). www.iso.org/iso/en/CatalogueDetailPage.CatalogueDetail?CSNUMBER=42546&scopelist=PROGRAMME

———. 2010. *ISO 26000 on Social Responsibility approved for release as Final Draft International Standard* (retrieved June 1, 2010). www.iso.org/iso/pressrelease.htm?refid=Ref1321

Johnson & Johnson. 2005. *2005 Sustainability Report*. www.jnj.com/community/environment/publications/2005_environ.pdf

Kaye, Leon. 2010. "Report from GRI Amsterdam: The Future of Transparent Sustainability Reporting." *Triplepundit*. June 1st, 2010 (retrieved June 2, 2010). www.triplepundit.com/2010/06/report-from-gri-amsterdam-the-future-of-transparent-sustainability-reporting/

Knight, Philip H. 2004. "Chairman's Letter to Shareholders: We're Faster Than Ever." *2004 Annual Report*. Oregon: Nike Inc.

Korean Occupational Safety and Health Agency. 2010. Enforcement *Decree of the Industrial Safety and Health Act* (retrieved June 1 2010). www.english.kosha.or.kr/main?act=VIEW&boardId=16&urlCode=T1||1240|1197|1197|1240|||/cms/board/board/Board.jsp&communityKey=B0488

L'Oreal SA. 2005. *2005 Sustainability Report* (retrieved August 10, 2006). www.loreal.com/_en/_ww/group/Img/LOREAL_RDD_GB.pdf

Locke, Richard M. 2005. *The Promise and Perils of Globalization: The Case of Nike* (retrieved January 15, 2005). www.mitsloan.mit.edu/50th/pdf/nikepaper.pdf

Moody, Andrew. 2009. "Multinationals Battling Locals for Market Share." *China Daily* (retrieved June 30, 2011). www.chinadaily.com.cn/business/2009-03/30/content_7628472.html

Mooney, Joseph W. 2004. "Investigate – Partner – then Export." *Northeast Export* (December), pp. 27–28.

Nash, James L. 2005. "Managing Global Safety: The Power of One." *Occupational Hazards* 67(9):28–32.

Nike Inc. 2003. *M.E.S.H.* (retrieved July 30, 2003). www.nike.com/nikebiz/nikebiz.html

———. 2005a. *FY 04 Corporate Responsibility Report*. www.nike.com/nikebiz/nikebiz.jhtml?page=29&item=fy04

———. 2005b. *Independent Monitoring & Assessment* (retrieved January 25, 2005). www.nike.com/nikebiz/nikebiz.jhtml?page=25&cat=monitoring&subcat=fla

Novartis AG. 2005. *Novartis Global Reporting Initiative 2005*. Basel, Switzerland: Novartis AG.

Seabrook, Kathy A. 1999a. "10 Strategies for Global Safety Management." *Occupational Hazards* 61(6):41–45.

———. 1999b. "Multinational Organizations." *Safety and Health Management Planning for the 21st Century*. James P. Kohn and Theodore S. Ferry, eds. Rockville, MD: Government Institutes.

———. 2006a. "A Briefing: Global Issues in Workplace Safety and Health." *World Focus. American Society of Safety Engineers International Practice Specialty*. 6(1):7–9.

———. 2006b. *An Interview with Stuart Wood, Administrator, International Practice Specialty, American Society of Safety Engineers*. Mendham, New Jersey: Global Solutions, Inc.

———. 2010. *2010 Global Safety and Health Briefing*. 2010 ASSE Professional Development Conference Proceedings, Session 534. Des Plaines, Il: American Society of Safety Engineers.

———. 2011. Global Environmental, Health and Safety Managemental Seminar. Presented at Seminarfest 2011, January 21–29, 2011. American Society of Safety Engineers, Las Vegas.

Secretaria de Segurança e Medicina do Trabalho. 2004. *Federal Law 6514/77 and Normas Regulamentadoras -5*. Sao Paulo, Brazil: Secretaria de Segurança e Medicina do Trabalho.

Singapore Ministry of Manpower and Workforce Development Agency. 2006. *Generic Occupational Safety and Health Competency Standards*. Singapore: Ministry of

Manpower and Singapore Workforce Development Agency.

The Economist. 2004. *Wood for the Trees* (retrieved May 29, 2005). www.economist.com

The White House. 2010. Office of the Press Secretary. May 22, 2010. "Press Release: Weekly Address: President Obama Establishes Bipartisan National Commission on the BP Deepwater Horizon Oil Spill and Offshore Drilling." (May 29 2010).

U.S. Census Bureau. 2005. *U.S. International Trade in Goods and Services Highlights* (retrieved May 29, 2005). www.census.gov/indicator/www/ustrade.html

Union Carbide Corporation. 2004a. *Chronology of Key Events Related to the Bhopal Incident* (retrieved November 18, 2004). www.bhopal.com/chrono.html

_____. 2004b. *Incident Review* (retrieved November 18, 2004). www.bhopal.com/review.html

_____. 2004c. *Opinion of the Attorney-General: Extradition of Warren Anderson* (retrieved November 18, 2004). www.bhopla.com/opinion.html

United Nations. 2006. *The Global Compact* (retrieved June 29, 2006). www.un.org/Depts/ptd/global.html

_____. 2010. KPMG Advisory N.V., United Nations Environment Programme, Global Reporting Initiative, Unit for Corporate Governance in Africa. "Carrots and Sticks—Promoting Transparency and Sustainability: An update on trends in Voluntary and Mandatory Approaches to Sustainability Reporting" (retrieved June 1, 2010). www.globalreporting.org/NR/rdonlyres/20F03459-4104-4B6D-AC3C-3C100F307EA2/4198/Carrrots2010final.pdf

United Technologies Corporation. 2004. *2004 Corporate Responsibility Data Index* (retrieved August 10, 2006). www.utc.com/responsibility_reports/2004/html/contents/page26.html

U.S. Congress. 2002. Sarbanes-Oxley Act of 2002, Public Law 107-204. 107th Cong., 2002.

U.S. Department of Labor. 2009a. *Top 10 Enforcement Citations*, January 30, 2009 (retrieved May 28 2010). www.osha.gov/dep/bp/Top_Ten_Enforcement.html

_____. 2009b. Occupational Safety and Health Administration. *Fact Sheet on BP 2009 Monitoring Inspection* (retrieved May 29, 2010). www.osha.gov/dep/bp/Fact_Sheet-BP_2009_Monitoring_Inspection.html

_____. 2009c. Occupational Safety and Health Administration. Press Release: October 30, 2009. "US Department of Labor's OSHA issues record-breaking fines to BP" (accessed May 31, 2010). www.osha.gov/pls/oshaweb/owadisp.show_document?p_table=NEWS_RELEASES&p_id=16674 09-1311-NAT

_____. 2010. Occupational Safety and Health Administration. "A Guide to Globally Harmonized System of Classification and Labelling of Chemicals" (retrieved February 23, 2010). www.osha.gov/dsg/hazcom/ghs

Vodafone Group Plc. 2006. *Group Corporate Responsibility Report for Financial Year 2005/06*. United Kingdom: Vodafone Group Plc.

APPENDIX A: ONLINE RESOURCES

Health and Safety Guidance and Regulations

General Resources

Global Solutions, Inc. Links to global websites sites with access to EHS regulations. www.globalEHS.com

International Network of Safety and Health Practitioner Organisations (INSHPO). Network of safety and health professional organizations founded in 2001. www.inshpo.org

Asia Pacific

Asia Pacific Occupational Safety and Health Organization(APOSHO) www.aposho.org

Australia

Australia National Occupational Health and Safety Commission. www.nohsc.gov.au/OHSLegal Obligations

Safety Institute of Australia. Australian association for safety and health professionals. www.sia.org.au

Baltic Region

Baltic Sea Network on Occupational Health and Safety. Provides regulations for Denmark, Estonia, Finland, Latvia, Lithuania Poland and Russia. www.balticseaosh.net

Canada

Canadian Center for Occupation Health and Safety. www.ccohs.ca

Canadian Society of Safety Engineers. Canadian association for safety and health professionals. www.csse.org

Health Canada. Information on the Workplace Hazardous Materials Information System (WHMIS)—Hazard Communication Requirements. www.hc-sc.gc.ca

European Union Countries

European Agency for Safety and Health at Work. www.agency.osha.eu.int

European Network of Safety and Health Professional Organizations (ENSHPO). www.enshpo.org

International Labour Organization (ILO). Provides United Nations conventions for worker safety and health. These conventions have been adopted by most countries in the World Trade Organization (WTO). www.ilo.org

India
Indoshnet. www.dgfasli.nic.in

Japan
Japan International Center for Occupational Safety and Health. Provides links to Web sites for regulations of other countries as well. www.jicosh.gr.jp/english/osh/list-of-laws.html

Malaysia
Laws of Malaysia - Act 514 and Occupational Safety and Health Act 1994. www.niosh.com.my/osha.htm

Korea
Korea Occupational Safety and Health Agency. www.kosha.or.kr/english/english.htm

Russia
Department of Occupational Health, Saint Petersburg Medical Academy of Postgraduate Studies. www.leivo.ru/mapo/model.html

Singapore
Ministry of Manpower. www.mom.gov.sg/MOM

United Kingdom
Health and Safety Executive. UK regulatory agency for worker health and safety. www.hse.gov.uk/pubns/hsc13.htm#2

Institution of Occupational Safety and Health (IOSH). British association for safety and health professionals. www.iosh.co.uk

United States
Board of Certified Safety Professionals. Registry of board-certified safety and health professionals. www.bcsp.org

Center for Disease Control and Prevention (CDC). Provides up-to-date medical advice for travelers around the globe. www.cdc.gov

Occupational Safety and Health Administration. Regulatory agency for worker health and safety. www.osha.org

United States State Department. Travel and visa information. www.travel.state.gov

Sustainability and Corporate Social Responsibility

Business for Social Responsibility. www.bsr.org
CERES HYPERLINK. www.ceres.org
Corporate Governance. www.corpgov.net/wordpress
CSR Wire. www.csrwire.com
Future 500. www.future500.org
Global Reporting Initiative. www.globalreporting.org/Home
International Organization for Standardization. *ISO 26000 Social Responsibility*. www.iso.org/iso/socialresponsibility.pdf
International Organization for Sustainable Development. www.iosd.org
SAM. www.sustainability-index.com/07_htmle/assessment/infosources.html

Supply-Chain Management Resources

Nike Sample Supply Chain Assessment Tools. www.nikebiz.com/responsibility/workers_and_factories.html

Avian Flu/Global Health Preparedness Resources

General
World Health Organization for Influenza Pandemic Preparedness Planning. www.who.int/entity/csr/disease/influenza/pandemic/en/index.html

Australia
"Fact Sheet – Pandemic Influenza Levels of Alert." www.health.gov.au/internet/wcms/publishing.nsf/Content/phd-health-emergency-threat-level.htm

General Information
www.health.gov.au/internet/wcms/publishing.nsf/Content/Pandemic+Influenza-1

European Union
Assessment Tool. National Pandemic Influenza Preparedness: Unit for Preparedness and Response at the European Centre for Disease Prevention and Control (ECDC) in Stockholm, in collaboration with the European Commission and the World Health Organisation Regional Office for Europe. www.ecdc.eu.int/Influenza/Assessment_Tool.php

United States
Checklist: Business Pandemic Influenza Planning. Department of Health and Human Services. www.pandemicflu.gov/plan/pdf/businesschecklist.pdf
United States Centers for Disease Control. www.cdc.gov/business

Appendix B: Case Study: The Union Carbide Bhopal India Incident

The following case study demonstrates the financial impact a U.S. multinational company can face when an incident that affects the safety and health of its employees and the surrounding community occurs at a non-U.S. site.

Background: The Incident

The following gives an overview as provided by Union Carbide Corporation. The purpose is to recognize how safety and health, operational, financial, and reputational risks can impact the survival of a business domestically and internationally.

On December 3, 1984, gas leaked from a tank of methyl isocyanate (MIC), resulting in a gas release at the Union Carbide plant in Bhopal, India. According to the state government of Madhya Pradesh, approximately 3800 people died and 11,000 people were left with disabilities (Browning 1993). Workers and their families living in close proximity to the plant site where they worked, which is the cultural norm in India, exacerbated the loss of life.

The Cause

The Indian government and Union Carbide confirmed the cause of the incident was the introduction of a large volume of water into an MIC tank, triggering a reaction that resulted in a gas release (Union Carbide Corporation 2004).

Reputational Risk and Financial Impact

Who were some of the first outside entities at the scene of the incident? The news media provided the world with a window to view the death and destruction caused by the incident. This event changed the future of Union Carbide in India. The reputation of Union Carbide was on the line; many questions, investigations, and litigation would befall the company for years following the explosion, and Union Carbide would sell its worldwide battery business in 1986 (Union Carbide Corporation 2004).

In February 1989, the Supreme Court of India directed Union Carbide Corporation (UCC) and Union Carbide India Limited (UCIL) to pay a total of $470 million in full settlement of all claims arising from the tragedy. In 1991, the Supreme Court of India also requested UCC and UCIL to voluntarily fund capital and operating costs of a hospital in Bhopal for eight years, estimated at approximately $17 million, to be built on land donated by the state government (Union Carbide Corporation 2004).

In November 1994, UCC completed the sale of its 50.9 percent interest in Union Carbide India Limited, a profitable operation for Union Carbide Corporation (Union Carbide Corporation 2004).

CEO Accountability

Another important factor in the Union Carbide incident is the implication of a CEO's criminal accountability for its global facilities when they are outside of where the global headquarters are domiciled. In the case of Union Carbide–Bhopal, the Attorney General of India, Soli Sorabjee, was asked to provide an opinion on whether extradition proceedings against Warren Anderson, Chairman of Union Carbide Corporation at the time of the gas tragedy, were legally sustainable (Union Carbide Corporation 2004). In the end, Mr. Anderson was not extradited. The premise for this verdict did not set a legal precedent for the criminal acts of a CEO, therefore, it leaves open the question of both criminal and civil legal liability of a CEO for the health and safety of workers and the public in India.

Mr. Sorabjee determined there must be the following "missing evidentiary links" (Union Carbide Corporation 2004):

1. the actual cause of the gas leak
2. Mr. Anderson's knowledge of the cause of the gas leak prior to its occurrence
3. the extent to which Mr. Anderson had decision-making control over UCIL's safety and design issues
4. whether Mr. Anderson refused to correct the hazard

Although the Attorney General of India determined that these missing evidentiary links (Union Carbide Corporation 2004) could be established, the Indian government never made a formal request for Mr. Anderson's extradition. According to the Union Carbide Website article, "Bhopal," general opinion by the United States' legal community was that, for humanitarian reasons, Mr. Anderson would probably not be extradited. He was 81 years old and in poor health, and there was a 17-year gap between the request for extradition and the Bhopal incident. Perhaps if Mr. Anderson had been younger and in better health, the Indian Attorney General, Soli Sorabjee, may have decided to pursue extradition (Union Carbide Corporation 2004).

The Bhopal facility incident impacted Union Carbide's reputation, along with its financial and legal standing in India. It also implicated the CEO, who could have been held legally responsible and accountable for operational incidents at Union Carbide sites throughout the world. That legal precedent was not tested. Today, with newer technology, news travels faster, images are more vivid, and information is transmitted without interruption. Organizations employ media management consultants who manage both positive and negative media events with a goal of making sure that the organization's reputation remains intact.

COST ANALYSIS AND BUDGETING

5

T. Michael Toole

LEARNING OBJECTIVES

- Be able to identify direct, indirect, and intangible costs and benefits that should be considered when evaluating safety investments.

- Be able to summarize and apply fundamental engineering economic concepts such as cash flows, time value of money, and discount rate.

- Be able to use common engineering economic methods, such as net present worth, annual cost, payback period, benefit-cost ratio, and internal rate of return.

ALL ORGANIZATIONS have goals. Most for-profit companies have corporate financial goals concerned with providing their shareholders with a certain rate of return on investment, which requires such companies to earn a certain level of profits, or net income. Within companies, resources such as people, equipment, physical plant, materials, and cash are *deployed* in ways thought to provide the best chances of meeting or exceeding corporate goals. Whether developing a new product, expanding into a new geographic area, or implementing a new information technology system to streamline operations, all projects require resources in order to bring the project to fruition, thereby eventually contributing to the achievement of corporate goals. One of management's key duties is to decide which projects—out of hundreds of possible projects—should be pursued. Managers must choose projects that they believe will deliver an acceptable *payoff* for the resources expended in pursuing them.

Profit is the difference between revenue and costs. Profit can be increased by increasing revenue, decreasing costs, or increasing revenues more than costs. Every safety professional knows that providing a safe and healthy work environment is a critical component in managing costs (ASSE 2002, Veltri et al. 2007). Although safety programs are rarely viewed as increasing revenues, they are typically recognized as essential to reducing costs, or to ensuring that costs do not increase faster than revenues do. Effective safety programs, therefore, can contribute to achieving corporate goals involving profitability just as much as more visible initiatives that involve new products, services, and so on.

In short, managers know that effective safety management is a good corporate investment. Like all investments, safety programs need dedicated resources. These resources usually include people, such as reassigned employees or newly hired employees, who must perform new safety functions. Resources needed for safety programs also routinely include infrastructure-related items such as office space, utilities, clerical support, and other administrative expenses. Specialized materials or equipment are also common safety-related resources.

Nearly all these resources require managers to spend money at the time the resources are received. In other words, the costs of resources are short-term costs. Yet, as is true of most strategic programs, the benefits or payoffs resulting from spending money now may not be felt until months, or even years, later. Moreover, because benefits occur in the future, they cannot be predicted with absolute certainty. Thus, managers who keep close watch over their organization's expenditures and profitability will be understandably reluctant to invest in any programs that may not deliver enough "bang for their buck." This may be especially true for safety programs, which have traditionally not been viewed as critical to company operations.

All managers—whether champions of safety or those with less progressive attitudes—need to be provided with thorough and objective analyses demonstrating that proposed safety expenditures represent solid investments deserving of their approval and unwavering support. Safety professionals know that the cost of preventing or reducing accidents is considerably less than the cost of actual accidents. But they need tools that can be applied to demonstrate that safety programs are cost-effective, worthwhile investments.

Several recent books and articles deal explicitly with this issue. In a book entitled *Safety and the Bottom Line*, Bird (1996) quoted the axiom of economic association: A manager will usually pay more attention to information when expressed or associated with cost terminology. Adams (2002) reported the results of a survey indicating that the need for "SH&E professionals to show management how safety can positively impact the bottom line" was widely recognized. Behm, Veltri, and Kleinsorge (2004) affirmed the need for making an economic case for safety, suggesting that framing safety investments using a cost-of-quality perspective might prove effective. In the book *Increasing Productivity and Profit through Health and Safety*, Oxenbaugh, Marlow, and Oxenbaugh (2004) also discuss the application of quantitative economic analysis to safety and health programs. Linhard (2005) discusses the development of a software called the "Return on Health, Safety and Environmental Investments" that analyzes the potential financial impacts of safety investments. Veltri et al. (2007) found statistically significant relationships between safety programs and operating performance in manufacturing firms. Specifically, the wider the gap between management and employee perceptions about safety programs, the higher the scrap and rework costs and the higher the production costs relative to competitors. Veltri and Ramsey (2009) provide a valuable review of the literature to identify the challenges of performing economic analysis and propose a detailed three-stage SH&E economic analysis model.

This chapter will provide the reader with tools that can objectively evaluate whether a project should be pursued. While the tools are often referred to as being associated with *engineering economics*, they are really financial analysis tools that are applied every day to projects having nothing to do with either engineering or safety. For example, the tools summarized in this chapter can help one decide whether to invest in real estate, launch a new product, or buy an extended warranty on a consumer purchase. Similarly, the fundamental concepts from engineering economics that will be introduced at the beginning of this chapter apply to all financial analysis situations, not just to those involving safety or engineering.

Fundamental Concepts Underlying Cost Analysis

Before introducing the powerful analytic methods that can be used to demonstrate that safety programs are good investments, it will be helpful to introduce several basic financial concepts and assumptions underlying these tools. This section will deal with life-cycle costing, direct and indirect costs and benefits, cash flow diagrams, time value of money, and the uncertainty of future costs and benefits.

Life-Cycle Costs

An important principle of engineering economic analysis is that all cash flows relevant to the decision must be included in the analysis. A cradle-to-grave approach should be taken, including all costs from the conception of a project (including engineering and design costs) through acquisition, startup, the end of its useful life, and disposal or salvage. The full life cycle of a project must be considered. Many consumers focus only on initial costs and ignore important future costs, such as in the purchase of computer printers, an instance in which many buyers are penny-wise but pound foolish, buying discount printers but paying many times their initial savings for expensive ink cartridges over the printer's life. The purchase of a car is another example of an instance in which some consumers do not consider costs accrued throughout the life cycle. Some buyers decide they can afford the higher initial cost of an expensive car but ignore the continuing need to purchase high-octane gasoline and pay more in insurance premiums and maintenance costs.

Life-cycle costing is also important for safety investments. Managers must think beyond the immediate and obvious costs and include in their analysis all annual maintenance costs and any end-of-life costs. Consider, for example, the hiring of a safety manager or other safety employee. The immediate and obvious cost is the employee's salary. Less obvious are the direct costs associated with payroll taxes, fringe benefits, new computer, and cell phone. Furthermore, annual increases in salary and overhead costs, the cost of periodic training, and similar costs should also be included. The analysis should also include any costs associated with the employee's termination (such as severance pay or unemployment) or retirement.

Safety-related equipment purchases also typically have obvious initial costs and less than obvious future costs. The large initial cash outlay in year 0 is followed by a series of annual benefits and costs. Annual benefits may include the operational cost savings per year resulting from the equipment and depreciation "charges." Annual costs often include maintenance or testing costs. At the end of the useful life of the equipment there may be a salvage benefit (as when used equipment is bought from the company). Alternatively, some equipment carries with it a required end-of-life disposal cost.

Indirect Costs and Benefits for Investment Decisions

A fundamental concept that applies to both short- and long-term projects is the concept of indirect costs and benefits. Most people would define a good project as one that results in more benefits than costs—a project that gives back more than is put into it. If asked to actually add up the benefits and costs, however, most only include direct benefits and costs, cash flows they consider to be directly associated with or directly resulting from a project that have obvious values. For example, OSHA's *Success Stories and Case Studies* Web page (2010) includes a link to success stories that summarize direct benefits.

It is important that financial analysis also include indirect costs and benefits and intangible costs and benefits (Labelle 2002). Some people use the terms tangible and intangible as synonymous with direct and indirect, but they are not. *Indirect costs and benefits* are tangible amounts that should be included in financial analysis but are not readily visible as being associated with an investment because one does not write a check for an indirect cost or cash a check for an indirect benefit. Indirect costs and benefits can be objectively quantified by people with expertise in cost accounting. Intangible costs and benefits are cash flows associated with an investment that are difficult to exactly apportion to the investment. They represent dollars that are real but are very difficult to quantify objectively, yet are still needed to perform the quantitative analysis discussed in this chapter. This subsection deals with indirect costs and benefits, and the next subsection addresses intangible costs and benefits.

Income Tax Benefits

Three specific types of indirect cash flows are summarized here. A full discussion of each of these cash flow types would require a book chapter alone, but they should at least be mentioned. One type of indirect cash flow is the effect of income tax. Investments are undertaken to generate profits, but these profits are subject to business income tax. Thus, proper cash flow diagrams for an income-producing investment should reflect the net income produced by the investment after tax. Because corporate tax rates are typically above 30 percent, income taxes can significantly reduce a potential project's apparent net income. Companies in special circumstances, allowing them to pay little or no corporate taxes, may not need to incorporate corporate taxes in their analyses.

Depreciation

A second important indirect cash flow that is easy to incorrectly omit is depreciation. Depreciation is an accounting concept, not an operational concept, but it carries with it implications for investment decisions. When certain types of major assets are purchased, their acquisition costs cannot be fully expensed (i.e., entered into the accounting books under cost of goods, thereby reducing net income) in the year they are purchased. Instead, the costs are amortized over the number of years set forth in tax law. In other words, although a firm may have paid a vendor the entire cost of a new piece of capital equipment, the income statement for that year can only include a portion of the costs. In each of the remaining years of the asset's service life, the income statement includes an entry for depreciation in the book value of the capital asset. This entry acts on paper like a current expense, thereby reducing net income and income tax owed. The U.S. government allows multiple methods of depreciation. Straight-line depreciation—in which the cost, less salvage value, of a purchase is simply divided by its useful life—is not commonly used, which is why depreciation is an indirect cash flow that frequently requires professional expertise to calculate.

Allocated Overhead Charges

The third type of indirect cash flow that should be mentioned is *allocated overhead* charges. Overhead costs may include portions of managers' salaries and fringe benefits, office space, utilities, and other costs that may be necessary infrastructure for a new investment. Adding overhead charges to cash flow diagrams typically reduces the attractiveness of a project because it reduces the positive cash flow associated with the investment. The amount of overhead to include in the analysis can be a difficult decision. Some companies establish a fixed rate and apply it to all direct costs. Other companies establish a fixed rate but often reduce or waive it during the early years of a new investment.

One reason it is sometimes difficult to determine whether overhead should be included in financial analysis is that overhead is typically associated with *fixed costs* (costs the company must pay regardless of the number of units produced, the number of hours worked, and so on) that may be independent of the investment under consideration. Existing management salaries and office rent and utilities usually are paid regardless of a new investment, so why should these costs be allocated to the investment under consideration? (The common-sense answer is that managers whose projects had to continue contributing toward corporate overhead would reasonably complain about other managers' projects that looked better because they did not have to contribute to overhead.) But to the extent that an investment ends up increasing fixed costs, the increases should clearly be assigned to the investment's project cash flow. *Variable costs* (costs incurred when additional hours are worked, etc.) should clearly be included in an investment's cash flow.

Intangible Costs and Benefits

Direct and indirect cash flows are those that are clearly associated with a project or investment and can be unequivocally measured. Intangible cash flows are those that are *real* but are nebulous and harder to quantify. ASSE's white paper on the return on investment for safety and health programs (ASSE 2002) refers to these costs as *hidden costs*.

Although there are always tangible costs associated with any safety program, there are rarely tangible benefits. Consider a company deciding whether to hire a full-time safety manager. One obvious set of costs would be the new manager's salary and fringe benefits. These are tangible costs because there are clearly future expenditures associated with hiring the new safety manager. However, if having a full-time safety manager was determined through appropriate analysis to automatically result in a 10 percent discount in workers' compensation or general liability insurance premiums, this would be reckoned a tangible benefit, because the savings would be easily quantifiable and would be directly associated with the new hire.

But such explicit benefits are often not present. Rather, the benefits associated with safety expenditures are usually intangible. For example, having a safety manager will eventually result in reductions in accidents, which will increase productivity and make the firm more attractive to clients who understand the importance of safety. Although these effects are important and may significantly improve the company's profits, they are hard to quantify, and it is difficult to *prove* as having resulted from the investment. If productivity increases, for example, line managers may attribute the increase solely to management initiatives wholly unrelated to safety. If other factors result in no increase in productivity, or even a decrease, line managers may fail to understand that productivity would have been even lower had the safety investment not been made.

It is important to note, however, that although intangible cash flows may be difficult to quantify and estimate, it is still usually appropriate to include them in economic analysis. After all, nearly all investment decisions are based on *estimates* of future returns on investments, not on locked-in cash flows. (This point is discussed more fully in the section entitled "Future Cash Flows Are Uncertain.") For example, Americans routinely buy equity stocks based on estimates of the future dividends and appreciation in the stock values. Product managers routinely decide to pay product development and marketing costs based on estimates of future sales and prices. Safety professionals should not hesitate to identify the specific intangible benefits associated with a safety investment, estimate their magnitudes, and let operational managers decide if these magnitudes should be adjusted or subjected to sensitivity analysis (a term that is discussed further in a section titled "Future Cash Flows Are Uncertain"). For example, many safety managers have historical data on the frequency of occurrence of specific types of injuries and the average direct cost associated with each type of injury. It is rational to estimate the savings resulting from an investment that will result in reduction in injuries by multiplying the number of fewer injuries expected by the average costs of each such injury.

Literature on Direct and Indirect Safety Costs

A number of authors have written about direct and indirect safety-related costs—especially on the indirect costs of accidents, which though often difficult to measure are nonetheless very significant. Leigh et al. (2000) reported that indirect costs comprise 71 percent of the total costs of an injury in the United States. It is important to point out, however, that this figure reflected all costs, including those borne by the injured employee and by government programs. An organization analyzing a safety investment, on the other hand, should only include costs borne by the organization, as explained below.

In an article titled "Hitting the Injury Iceberg," Brandt (1999) stated that "most insurance loss control experts agree the indirect costs of workplace injuries (costs usually not covered by insurance) can range anywhere from two to ten times the cost of the face value of the claim." Kinn et al. (2000) reported that indirect costs may range from two to twenty times the direct costs in the cases of plumbers and pipefitters. The introduction to OSHA's Safety & Health Management Systems eTool states that "studies show that the ratio of indirect costs to direct costs varies widely, from a high of 20:1 to a low of 1:1" (OSHA 2010).

Brody, Letourneau, and Poirer (1990) identified indirect accident costs for general industry, including wages paid to employees at the accident site immediately after an accident (which are not covered by insurance), damaged materials, occupation of administrators' time, production losses, tarnished public image, deteriorated labor relations and morale, and higher wages (due to perceived risk). This same article also identified prevention costs, including fixed prevention costs (those that occur before production occurs and do not vary with accident rate), variable prevention costs (those that occur as a result of accidents in an effort to prevent recurrences, such as accident investigations), and unexpected prevention costs (those associated with reducing the risk of injury but not foreseen at the time a piece of equipment was purchased).

Brady et al. (1997) identified (1) the direct costs associated with a specific injury, (2) the indirect costs associated with a specific injury, and (3) the indirect costs associated with other health and safety requirements (such as prevention programs) instead of with a specific injury (see the lists set out below). Lanoie and Tavenas (1996) reported on a detailed study of the direct and indirect costs and benefits of a participatory ergonomics program.

Direct Costs Related to Specific Illness or Injury

Direct costs related to a specific illness or injury include the following:

- medical care
- physician, other provider services
  - inpatient
  - outpatient
- clinic, hospital services
  - inpatient
  - outpatient
- ancillary diagnostic services
  - laboratory
  - radiology
  - electrocardiography
  - electromyography
  - other
- patient-specific medical supplies, equipment
- medications, pharmacy services
- rehabilitation, occupational/physical therapy
- employee assistance counseling
- other injury- or illness-specific requirements
- workers' compensation payments (as applicable)
- sick pay (as applicable)
- other benefits (as applicable)
- compliance with Occupational Safety and Health Administration reporting required for the illness or injury
- costs for temporary employee to accomplish tasks of ill or injured worker
- case-management costs
- vocational rehabilitation counseling
- case-specific litigation costs
- case-specific human resources or personnel costs
- specific accommodations required by the Americans with Disabilities Act

Indirect Costs Related to Specific Illness or Injury

Indirect costs related to a specific illness or injury include the following:

- reduced productivity
  - absence of injured or ill employee
  - shift in activities of co-worker to accomplish absent employee's work
  - increased supervisor effort to cope with absence of employee
  - temporary or long-term absence of corporate memory possessed by the ill or injured employee
  - start-up and training time for replacement employee
  - start-up and training time for the returned ill or injured employee
  - development of limited work position for ill or injured worker, as appropriate
  - reduced effectiveness of "nearby" co-workers
  - overtime pay
- impacts on competitiveness
  - potential for reduced customer satisfaction because of absent employee and effects of employee's absence
  - effect of greater-than-projected medical costs
  - increased risk of illness or injury by temporary or replacement employee because of limited time for hazard or safety training, or other factors
  - increased insurance premiums
  - increased overtime costs
  - increased training and retraining costs
  - increased legal costs, including class-action defense, coordination of new policies to respond to event or prevent recurrence, and other such costs
  - loss of senior management time in responding to event
  - reduced performance and effectiveness of ill or injured person once returned
  - effects on labor relations, including potential strikes as well as requests for hazard

pay and new safety programs or protective equipment
- potential for adverse media coverage
- effect on worker morale (which also impacts productivity)
- requirements for increased quality-control efforts, as required, for replacement or returning employee
- increased human resources and personnel department costs incurred in efforts to replace ill or injured worker
- medical, industrial hygiene, and safety costs involved in investigation of accident or exposure site
- risk-management activities involved in investigation of accident or exposure site, or other activities

Indirect Costs Related to Other Health and Safety Requirements, Not to a Specific Illness or Injury

Indirect costs related to other organizational health and safety requirements include the following:

- the health and safety program (in-house, consultant, contract) costs
  - staffing to provide care for ill and injured employee on site (or costs to arrange such capability off site)
  - regulatory compliance, including medical/industrial hygiene monitoring, surveillance programs
  - development and maintenance of capabilities for case management
  - employee assistance program other than costs for specific illnesses or injuries
  - development of health and safety policies
  - data-processing and data-management costs
  - research expenditures, as appropriate
  - wellness, health promotion, immunizations
  - health and safety committees
  - evaluation of options for provision of services
  - program evaluations
  - interactions with other organizations, departments, and managers, including risk managers
  - drug and alcohol testing programs
  - preplacement and periodic examinations and evaluations
  - other organization-specific costs
- other costs of health and safety activities—other than those involving a specific illness or injury
  - human resources and personnel
  - benefits
  - legal
  - labor relations and unions
  - management (other than health and safety)
  - other organization-specific costs

Hinze (2000) investigated typical indirect costs associated with construction injuries and reported that indirect costs are typically four times direct costs. He identified the following typical indirect costs associated with medical case injuries in construction:

- 15.7 hours lost time by the worker
- 3 hours transporting the worker from the job site
- 12 hours lost by the injured worker's crew
- 4 hours to repair damaged equipment and material
- 5.5 hours lost by the supervisor responding to and reporting the accident

Hinze also identified typical direct costs associated with the following common safety investments for construction firms:

- substance abuse testing
- safety inspectors and managers
- safety training
- personal protective equipment (PPE)
- safety committees
- accident investigations
- safety incentives

Behm, Veltri, and Kleinsorge (2004) provided a helpful framework for analyzing direct and indirect costs. Specifically, they discussed how a cost-of-safety analysis should include four classes of costs and benefits: prevention, detection, internal failure, and external failure. They suggest that the appropriate level of spending on a safety project or total safety

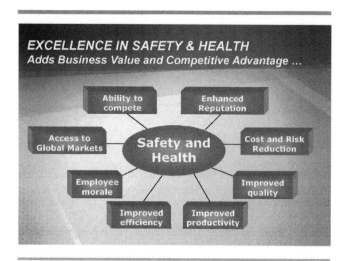

FIGURE 1. Indirect benefits of a safety and health program (OSHA et al 2005)

program is the point where the sum of prevention and detection costs equals the sum of internal and external failure costs. Their article also articulates the practical challenges of collecting the necessary data.

OSHA has made available an eTool called "$afety Pays" (OSHA 2010), which calculates both direct and indirect cost savings, and the resulting increase in profit margins, from estimated reductions in specific types of injuries. OSHA has also made available a powerpoint file (OSHA et al. 2005) that includes slides relating to indirect benefits (see Figure 1) and indirect costs. NIOSH (2004) has made available a practical guide for collecting and evaluating direct, indirect, and intangible costs and benefits associated with safety investments under consideration. The process recommendations are summarized in the opening chapter of this handbook.

Cash Flow Diagrams

One of the first steps in performing an effective evaluation of any investment should be creating a cash flow diagram (CFD) of the proposed investment. As the name implies, CFDs are diagrams that indicate all of the incoming and outgoing monies associated with an investment. CFDs are two-dimensional and have an invisible x–y grid underlying them. The x-axis represents time and typically shows units of months or years. The y-axis is the amount (that is, magnitude) of the cash flow and typically shows units in dollars or thousands of dollars.

The first step in creating a CFD is to identify the length of time associated with the investment decision. The length of time is typically the life cycle of the item or project being considered. For example, a piece of capital equipment might have an expected service life of 25 years, but PPE might have an expected service life of only five years. At the end of the investment's service life the item must be replaced, or the project repeated, with the same cash flow assumed during the first life cycle. Because the cash flows associated with the subsequent items or projects is assumed to be identical to those of the first item or project, there is no need to repeat them again. The evaluation can be made solely on the cash flows expected during the first service life. Once the length of analysis is determined, a scaled horizontal line 2 to 4 inches long is drawn with tick marks for each month or year, as shown by the horizontal line in Figure 2, which is a CFD for typical capital equipment. Each tick mark represents the end of that period. For example, the tick mark labeled "2" represents the end of year 2.

The next step is to draw vertical arrows corresponding to each cash inflow (benefit) or outflow expected each year. Note that a cash flow diagram is always created for the perspective of one individual or organization. Different entities associated with a transaction may have very different cash flow diagrams for the same transaction. Cash outflow lines are drawn with the arrows pointing down, and inflows

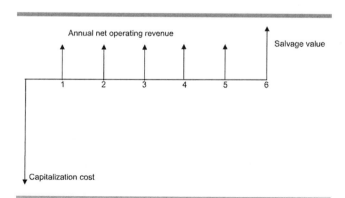

FIGURE 2. Typical cash flow diagram for capital equipment

up. The length of each arrow is scaled to provide a graphical indication of the magnitude of the cash flow it represents. Thus, a $2000 outflow should be twice as long as a $1000 outflow. The amount associated with each arrow is usually shown near the arrowhead. It is traditional to show the total amount of inflows or outflows in a given period at the end of that period. Thus, all cash flows expected during the second year of the investment would be shown by an arrow located at the tick mark labeled "2."

Time Value of Money

One of the assumptions underlying investment decisions, the Western banking system, the stock and bond markets, and essentially all economic systems, is that there is a time value of money. The time value of money refers to the idea that having a certain amount of money now is not the same as having an identical amount of money a few months or years from now. Everyone recognizes this time value of money even without using the term. Someone offered $100 would be happy. If asked whether he wanted it now or a year from now, he would be forced to consider the various pros and cons of having $100 now versus having it a year from now.

Part of the preference reflects the old saying that "a bird in the hand is worth two in the bush." If not given the money now, there is a chance that he will never receive it. But an even more quantifiable reason why one might prefer the $100 now is that he could spend it right away or put it in a savings account at a bank and know that one year from now the money would have accrued interest (money paid by a borrower, such as a bank, for the use of others' money). So if one receives $100 now, he could have more than $100 a year from now. On the other hand, if one does not receive $100 until a year from now, he would clearly only have $100 then.

Opportunity Costs

The simple example above illustrates an important concept related to the time value of money: *opportunity cost*. Money (and any other resource) can be invested today to possibly secure even more tomorrow. The notion of opportunity cost is important whether one is the lender or the receiver. If a person loans someone $100, he cannot earn bank interest on that money until he is paid back, so he might likely charge the borrower interest to compensate for the bank interest he would otherwise have earned. As a borrower, he recognizes that the lender could have used the money loaned out for other things, so he is willing to return interest for the privilege of having the money. However, the return the lender hopes to receive from the money must outweigh the interest willing to be paid. It would not make sense to borrow money at 10 percent interest so that one could put it in a bank account where it would earn 5 percent interest.

The concept of opportunity costs applies to companies just as much as—if not more than—to individuals. For-profit companies exist for the explicit purpose of earning profits that allow those companies to provide satisfactory returns on investments to their shareholders. Companies may establish many company goals each year, but goals expected to increase net income always top the list. Large and established companies typically have formal, explicit procedures for evaluating new expenditures to increase the chances that all monies spent will increase profit margins. Established companies do this by establishing minimum rates of return or discount rates. The term *discount rate* reflects the fact that all future expenditures or benefits are discounted (or systematically reduced) by an interest rate to be used in project evaluations. The choice of discount rate is discussed later in the chapter.

Equivalency

The concept of a discount rate can be better understood by introducing another new term, *equivalency*. As a simple example, assume that one is offered either $1000 now or $1060 one year from now; and assume that the money is received now and left untouched in a bank account earning 6 percent interest. Either way, one ends up with $1060 one year from now. The two cash flows are considered equivalent. Although they seem to involve different amounts of money ($1000 versus $1060), once the time value of money is considered they are recognized as equivalent.

The formula underlying the time value of money and equivalency is provided in Equation 1.

$$F = P(1 + i)^n \qquad (1)$$

TABLE 1

Compound Interest Formulas

Present Worth Factors $(1+i)^n$

Number of Periods	Interest rate									
	1%	2%	3%	4%	5%	6%	7%	8%	9%	10%
1	0.990	0.980	0.971	0.962	0.952	0.943	0.935	0.926	0.917	0.909
2	0.980	0.961	0.943	0.925	0.907	0.890	0.873	0.857	0.842	0.826
3	0.971	0.942	0.915	0.889	0.864	0.840	0.816	0.794	0.772	0.751
4	0.961	0.924	0.888	0.855	0.823	0.792	0.763	0.735	0.708	0.683
5	0.951	0.906	0.863	0.822	0.784	0.747	0.713	0.681	0.650	0.621
6	0.942	0.888	0.837	0.790	0.746	0.705	0.666	0.630	0.596	0.564
7	0.933	0.871	0.813	0.760	0.711	0.665	0.623	0.583	0.547	0.513
8	0.923	0.853	0.789	0.731	0.677	0.627	0.582	0.540	0.502	0.467
9	0.914	0.837	0.766	0.703	0.645	0.592	0.544	0.500	0.460	0.424
10	0.905	0.820	0.744	0.676	0.614	0.558	0.508	0.463	0.422	0.386
11	0.896	0.804	0.722	0.650	0.585	0.527	0.475	0.429	0.388	0.350
12	0.887	0.788	0.701	0.625	0.557	0.497	0.444	0.397	0.356	0.319
13	0.879	0.773	0.681	0.601	0.530	0.469	0.415	0.368	0.326	0.290
14	0.870	0.758	0.661	0.577	0.505	0.442	0.388	0.340	0.299	0.263
15	0.861	0.743	0.642	0.555	0.481	0.417	0.362	0.315	0.275	0.239
16	0.853	0.728	0.623	0.534	0.458	0.394	0.339	0.292	0.252	0.218
17	0.844	0.714	0.605	0.513	0.436	0.371	0.317	0.270	0.231	0.198
18	0.836	0.700	0.587	0.494	0.416	0.350	0.296	0.250	0.212	0.180
19	0.828	0.686	0.570	0.475	0.396	0.331	0.277	0.232	0.194	0.164
20	0.820	0.673	0.554	0.456	0.377	0.312	0.258	0.215	0.178	0.149
21	0.811	0.660	0.538	0.439	0.359	0.294	0.242	0.199	0.164	0.135
22	0.803	0.647	0.522	0.422	0.342	0.278	0.226	0.184	0.150	0.123
23	0.795	0.634	0.507	0.406	0.326	0.262	0.211	0.170	0.138	0.112
24	0.788	0.622	0.492	0.390	0.310	0.247	0.197	0.158	0.126	0.102
25	0.780	0.610	0.478	0.375	0.295	0.233	0.184	0.146	0.116	0.092

In this equation, F is the amount one will end up with after n years if earning i percent interest on an initial amount P. This equation may initially seem a bit intimidating, but it is very intuitive. If one starts with $1000 and can earn 6 percent interest yearly, he will have $1060 after one year: the original $1000 plus $60 in interest [1000 + (1000 × 0.06) = $1060] during the year. Thus the equation for the amount after one year is simply $P(1 + i)$. During the second year, because interest is earned on the amount one started with at the beginning of the year [$P(1 + i)$], one ends up with $P(1 + i)(1 + i)$, which can be written more simply as $P(1 + i)^2$. If this logic is continued, it is not hard to see that the amount accrued after n years is $P(1 + i)^n$. (*Simple interest* arrangements, in which interest is not earned on previously earned interest, will be discussed shortly.)

This equation can also be used in reverse to calculate how much should be invested now (P) in order to end up with F after n years if the interest rate is i percent. This equation defines equivalent cash flows. That is, if the interest rate of i percent is assumed, having P now or F in n years from now are equivalent.

The term $(1 + i)^n$ is referred to as the *single payment discount factor*—written as $(F/P, i, n)$ and read as "Find F given P, i, and n." Because not every calculator has a y^x button, most engineering economics textbooks include the single payment discount factors for different combinations of interest rates (i) and number of time periods (n). Table 1, which was created using a spreadsheet, provides an example. Such tables make it easy to multiply P by the discount factor to calculate F or to divide F by the discount factor to calculate P.

It is important to note that individual cash flows are additive. To calculate how much money one would need to have now if he were going to pay out $2000 at the end of year 3 and $3000 at the end of year 5, simply multiply the amount of each payment by the appropriate discount factor and add the products together. $P = (2000)(0.840) + (3000)(0.748) = \$3921$ for $i = 0.06$ (see Table 1).

Calculating F or P if the cash flow diagram includes a series of uniform cash flows can be accomplished through two methods. One is to follow the procedure in the previous paragraph and add up the products of each cash flow and the appropriate discount factor. An easier way is to use Equation 2:

$$P = A \frac{(1+i)^n - 1}{i(1+i)^n} \qquad (2)$$

The A in Equation 2 is the amount of the uniform payments (or cash outflows), which are assumed to start in the first year and occur each year through year n. The portion of the equation to the right of the A is the uniform series discount factor, which is abbreviated as $(P/A, i, n)$ and is often tabulated for various combinations of interest rates and number of time periods. Such tables make it relatively easy to calculate how much one would need now to be able to make uniform payments over a specified number of time periods. For example, because the discount factor for a uniform payment over five time periods at an interest rate of 6 percent per time period is 4.212, one would need $4212 now to make these payments of $1000 each year.

Arithmetic and Geometric Gradients

There are two additional sets of discount factors that are less common than the single payment or uniform series factors discussed above: *arithmetic gradient* and *geometric gradient*. As the names suggest, both factors reflect gradients, or increasing patterns. Arithmetic gradients increase linearly, such as by 10 percent each time period. (For example, an analysis could include health-insurance costs, which have historically increased by ranges of percentages above inflation.) The equation for calculating the present worth of a cash flow with an arithmetic gradient is provided in Equation 3. Note that these equations assume there is no cash flow during the first year and that the cash flow is G in year 2, $2G$ in year 3, and so on, and $(n-1)G$ in year n. Geometric gradients increase nonlinearly. The equation for calculating the present worth of a cash flow with a geometric gradient is provided in Equation 4. Readers who need the equations for gradient series discount factors should refer to Blank and Tarquin (2005), to Newnan, Eschenbach, and Lavelle (2004), or to other engineering economic textbooks.

$$P = G \frac{(1+i)^n - in - 1}{i^2(1+i)^n} \qquad (3)$$

$$P = A_1 \frac{1 - (1+g)^n(1+i)^{-n}}{1-g} \qquad (4)$$

Instead of looking up the appropriate discount factor in a published table or punching the formula into a calculator, one can use financial analysis functions built into most spreadsheets. In *Microsoft Excel*, for example, financial functions include *PV* (present value), *FV* (future value), *PMT* (uniform payment), *NPV* (net present value), and others.

Effective Interest Rate versus Nominal Rate

It is important to note that economic analysis should be based on the *effective interest rate*, not on the *nominal rate*. The nominal rate is typically given in years (as for bank accounts) or months (as for credit cards). The effective rate reflects how often during the nominal time period interest is actually calculated. For example, bank savings accounts are often quoted as "x percent annual interest, compounded quarterly." This means that each quarter, the appropriate prorated interest rate is applied to the account's balance at that time. The principal that earns interest during the second quarter is therefore greater than if interest had not been calculated until the end of the year.

As an example, compare the balance in an account if $1000 is deposited for three years in a savings account earning 6 percent interest compounded annually instead of quarterly. The single payment discount factors found in tables in the back of engineering economic textbooks make this comparison simple. The discount factor for the account compounded annually is $(F/P, 6\%, n = 3) = 1.191$. The discount factor for the account compounded quarterly is $(F/P, 1.5\%, n = 12) = 1.196$. In this case, the account compounded quarterly would earn $5.00 during three years, illustrating that when the compounding period is less than a year, it is necessary to divide the nominal rate by the number of compounding periods per year and multiply the number of years by the number of compounding periods per year when using compound interest tables.

The differences between nominal and effective interest rates are related to the differences between simple and compound interest. *Simple interest* is the amount of interest earned with no intermediate compounding. If one borrows $1000 at a 6 percent interest rate for three years, the interest payment at the end of three years is $180 (1000 × 0.06 × 3) under a simple interest arrangement. *Compound interest*, on the other hand, adds interest to the principal at the end of each compounding period. If one borrows $1000 at 6 percent interest for three years, the interest payment at the end of three years is $191 under a quarterly compounded interest arrangement. (The previous paragraph shows that the future value factor was 1.191 for this combination of interest rate and number of time periods).

Future Cash Flows Are Uncertain

The examples used thus far have been simple because they involved interest charged for loans or earned in bank accounts. Interest rates are usually fixed, which means that the cash flows associated with interest payments are precisely known ahead of time. When taking out a car loan, the purchaser knows he will have to pay back a certain amount of money each month (a sum including both a portion of the principal and interest on that portion). Cash flows associated with most investment decisions, however, are not nearly so predictable. The exact amount of cash flows occurring right away are known, but the exact amounts of cash flows in future periods are unknown.

Immediate and Future Cash Flows

But even buying a new car is not as predictable as one might think. When attempting to draw a cash flow diagram for this decision, all immediate and future cash flows should be included. The purchaser should know exactly how large the initial downpayment will be as well as the exact amount of the monthly car-loan payment. But the cash flow diagram should also include the amounts needed for car insurance, maintenance, gasoline, and similar considerations. Gasoline costs can be predicted with a high degree of certainty during the first year, but costs in future years are less certain: one's travel habits could change, the price of gas may escalate, and so on. Similarly, future maintenance or insurance costs cannot be reliably predicted.

The simple example above illustrates that specific amounts of future cash flows cannot be predicted with certainty unless they are associated with a fixed interest rate or with some type of business contract. Insurance premiums are a form of contract. In the case of health insurance, for example, a firm agrees to pay a specified premium each month, and the health insurance company agrees to pay for the actual costs of providing the firm's employees with major medical healthcare, whether costs are below or above the total premiums. (Employee co-pays and major medical caps are ignored here for simplicity's sake.) But a company's contract with a health insurer typically lasts a year, or less. At the end of that year, the company will have the choice of signing a new contract that will likely include higher premiums. In other words, businesses have no control over price escalation.

Most cash flows associated with investment decisions cannot be fixed through a contract. Going back to the example of hiring a safety manager, even the tangible costs associated with this decision are subject to a high degree of uncertainty. The employer could know with confidence the amounts of salary, fringe benefits, and training costs this person would receive the first year she worked for the company, but the exact amounts for these cost categories are less certain for the second year and even more uncertain for later years. Similarly, even if the employer felt he could quantify the intangible benefits of hiring this person during the first year (reduced accident costs, increased productivity, and other such things), the estimated benefits in future years would be considerably less certain.

Threat of Inflation

Even were it possible to know the benefits of an investment for several years into the future, the threat of *inflation* would still cause the real value of one's future cash flows to be uncertain. Inflation is the economic phenomenon by which labor wages and the prices of most goods rise—which means that a given sum of money will buy less a year from now than it does today. If $50,000 today purchases 50 sets of PPE, $50,000 might only purchase 40 sets of PPE a year from now, if inflation is excessive. Thus, even if it were somehow possible to calculate the amount of cash one will, without question, receive from an investment each year, controlling the actual worth of those dollars' purchasing power is impossible. Thus, inflation will always decrease the actual rate of return for investments involving a sizable initial expense followed by a stream of positive cash flows (such as benefits), because the future positive cash flows will be worth less in terms of real dollars because of inflation. Inflation can also pose an investment risk if benefits inflate at a different rate than costs do. For example, if the benefits of a safety investment included reduced workers' compensation insurance premiums and the costs included a safety manager's salary, the payback in real dollars would be less than the payback in actual dollars if the salary inflated at a greater rate than the insurance premiums did.

Because future cash flows cannot be predicted with certainty, two things should be done when making investment decisions. First, all future cash flows should be discounted, or multiplied by a factor that makes their present worth well below their future nominal amounts. The interest rate or discount rate used in the analysis reflects both the uncertainty of the future cash flows and the opportunity cost. The higher the uncertainty of the future cash flows, the higher the discount rate.

Sensitivity Analysis

Next, sensitivity analysis should be performed. This involves identifying the assumptions underlying the analysis, changing those assumptions, and rerunning the analysis—analyzing how sensitive the outcome is to variation in the assumed cash flows. An investment that looks worthwhile under one set of assumptions may be very unattractive under another set of assumptions. If the likelihood that the second set of assumptions will turn out to be true is considerable, it should not be ignored. Proper risk-management principles require a manager to at least consider other possible scenarios, and sensitivity analysis is a tool that does this thoroughly and systematically.

For example, when analyzing whether to hire a safety manager, one might run a series of analyses with different rates of escalations in healthcare insurance, with different amounts of productivity increases, and with different discount rates (because the time value of money for the company may change if the prime rate changes). Because sensitivity analysis typically involves a lot of number crunching, it is easiest to enter the estimated cash flows into a spreadsheet, then methodically vary each assumed value individually, noting how each change affects the final value. Table 2 provides an example in which the net result on internal rate of return (discussed shortly) of an investment ranges between 4 and 12 percent, depending

TABLE 2

Sensitivity Analysis			
Health Insurance Increase	Productivity Increase	Discount Rate	Internal Rate of Return
8	1	5	5.2%
10	1	5	5.5%
12	1	5	5.7%
8	2	5	7.8%
10	2	5	8.3%
12	2	5	8.8%
8	3	5	10.4%
10	3	5	11.2%
12	3	5	12.0%
8	1	7	4.6%
10	1	7	4.8%
12	1	7	5.1%
8	2	7	6.9%
10	2	7	7.4%
12	2	7	7.8%
8	3	7	9.2%
10	3	7	9.9%
12	3	7	10.6%
8	1	9	4.0%
10	1	9	4.2%
12	1	9	4.4%
8	2	9	6.0%
10	2	9	6.4%
12	2	9	6.8%
8	3	9	8.0%
10	3	9	8.6%
12	3	9	9.2%

on whether healthcare insurance is assumed to increase by 8, 10, or 12 percent, productivity is assumed to increase by 1, 2, or 3 percent, and the discount rate is assumed to be 5, 7, or 9 percent. By estimating the likelihood of each possible situation (for example, that there is a 20 percent chance that healthcare will increase 12 percent per year), one can determine the risk that the actual return on investment will turn out considerably less than expected. If performing a manual sensitivity analysis such as the one in Table 2 seems somewhat tedious, one may wish to consider using the Scenarios tool in *Microsoft Excel*.

Expected Value

The previous paragraphs have indicated that most cash flows and other outcomes associated with safety investments cannot be predicted with certainty and that effective managers will consider the possible outcomes of their decisions. This discussion emphasizes the risk that underlies all management decisions. Managers make decisions based on the predicted outcomes of their chosen courses of action. The fact that the actual outcomes may vary significantly from those predicted means that the decision may appear, in hindsight, to have been the wrong choice. The final fundamental concept is *expected value*, a rational method for making risky decisions.

The expected value of a choice is the value of the sum of each possible outcome multiplied by the probability associated with each outcome. For example, if a friend offers to flip a coin and the winner pays the other $1, the expected outcome would be the probability of a head (0.5, or 50 percent) multiplied by the outcome of a head (winning $1) plus the probability of a tail (50 percent) multiplied by the outcome of a tail (losing $1). The expected value of this bet therefore equals (0.5)(+1) + (0.5)(–1), which equals zero, meaning that one should not really care if he takes the bet or not. One might be ahead or behind if he took this bet several times, but over the long run, the heads and tails would balance out and one would end up even.

The next calculation is for the expected value of a slightly more complicated choice: whether to hire a safety inspector. (Note that this analysis will be grossly

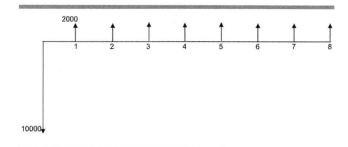

FIGURE 3. Payback-period example

oversimplified for this example.) Assuming that the company will definitely incur total costs of $100,000 per year for salary, fringe benefits, office space, and other such expenses, an employer estimates there is a 25 percent probability that this person will be totally ineffective and result in no benefit at all to the company; a 50 percent probability that this person will result in productivity savings of $100,000; and a 25 percent probability that the inspector will make a bigger difference, resulting in a total savings of $250,000. Should the inspector be hired? A risk-averse manager might focus on the admittedly significant chance (25 percent) that the person may be a complete loss, costing the company $100,000, but a smart manager will recognize that the expected value of this inspector is positive: (1)(–100,000) + (0.25)(0) + (0.5)(100,000) + (0.25)(250,000) = $12,500. Because this number is significantly positive, the inspector should be hired.

Real-world decisions are much more complicated than the oversimplified example above, but expected value still represents a tool that managers can use to make rational decisions. Managers often make irrational choices because they focus excessively on one aspect of a decision, such as the risk that the worst possible outcome may occur. Expected value can be a helpful tool for forcing managers to methodically consider *all* outcomes—not just the ones that may make managers look particularly good or exceptionally bad.

INVESTMENT ANALYSIS METHODS

Now that the fundamental concepts underlying engineering economics and financial analysis have been presented, it is time to introduce the specific analytical techniques that can be applied to make effective deci-

Cost Analysis and Budgeting

sions about safety investments. As stated at the beginning of this chapter, these concepts and techniques can be appropriately applied to nearly all investment decisions, whether safety management is involved or not.

Payback Period

The simplest but least-effective analytical technique is *payback period*. The payback-period procedure involves simply determining the earliest point in time when cash outflows are exceeded by cash inflows. Assume, for example, a safety project that cost $10,000 to implement in year 0 and provided benefits of $2000 each year indefinitely. The cash flow diagram for this project is shown in Figure 3. The payback period for this investment would be five years because, by the end of the fifth year, the initial outlay would have been offset by the cumulative annual savings.

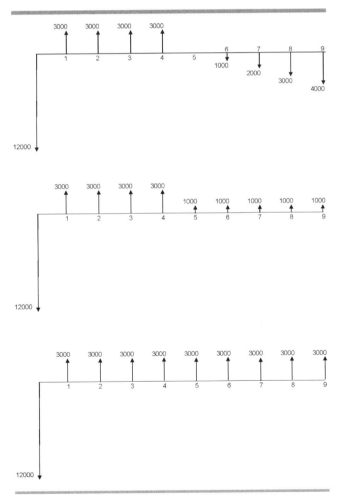

FIGURE 4. Illustration of payback period's cash flow projection shortcomings

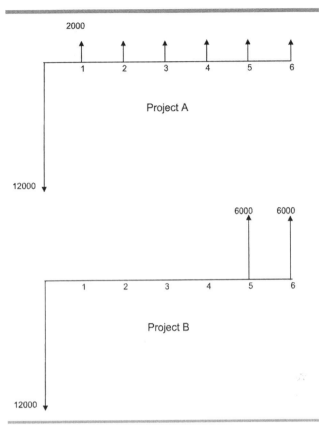

FIGURE 5. Illustration of payback period's benefit-cost shortcomings

The primary strength of the payback-period method is its simplicity. One hardly even needs a calculator to do the math. Another strength is that it captures the question that some managers first ask about a candidate investment: "How soon before I get my money back?"

The weaknesses of the payback period are severe. Most fundamentally, it ignores all cash flows after the time when total cash inflows equal total cash outflows. Consider the three cash flow diagrams shown in Figure 4, all of which are from the perspective of the same firm. All three projects have a payback period of four years, but the cash flows that occur after year 4 are clearly very different. Project A has a very negative cash flow after year 4, Project B has a decent cash flow after year 4, and Project C has a very attractive net cash flow after year 4. Yet the payback-period approach would rate all three projects as being equally attractive.

Another major weakness of the payback period is that it ignores the time value of money. Consider, for example, the two cash flow diagrams in Figure 5.

Both projects have a payback period of 6 years, but Project A's cash inflows start immediately while Project B's cash inflows do not occur until year 5. Clearly, the early cash inflows from Project A could be invested right away to earn interest or other types of earnings, making it more attractive than Project B.

A third weakness of the payback period is that it does not indicate how attractive a project is. Projects A and B in Figure 6 both have the same payback period, but Project B is clearly more attractive than Project A because the amount of cash inflow in the fifth year is twice that of Project A.

Break-Even Analysis

Break-even analysis is in some ways similar to payback period and in other ways is quite dissimilar. It resembles payback period because both methods involve determining a point where the incremental benefits (periodic cash inflows) first offset the initial costs (cash outflows). However, when reckoning with payback period in mind, the initial costs are the capitalization costs, and the cash inflows are associated with each time period. When reckoning with break-even analysis, the initial costs tend to be fixed costs, and the incremental benefits are variable cash inflows associated with each unit sold, or are otherwise associated with the initial costs.

The decision about whether to hire a full-time safety inspector could serve as an example. The new hire's salary and fringe benefits are clearly fixed costs associated with the decision. To the extent that the mere circumstance of having a full-time safety inspector resulted in a certain number of additional sales to clients who value safety, one set of benefits accruing from this hire would be fixed. On the other hand, another set of benefits accruing from having this employee would likely be variable, because each labor hour worked by the company should have a lower chance of an accident and, therefore, eventually result in a lower workers' compensation insurance premium. An appropriate break-even analysis should calculate the direct labor payroll the company needed to offset the safety inspectors' fixed costs, less fixed benefits.

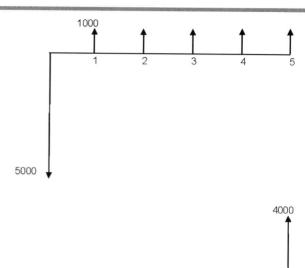

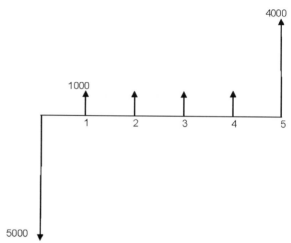

FIGURE 6. Illustration of the payback period's shortcomings when measuring project attractiveness

Assume that the total fixed costs associated with having a safety inspector are $100,000; that simply having an inspector results in $40,000 of additional profit; and that the workers' compensation reduction, per $100 of payroll, is $25. The break-even analysis would therefore be $0 = -\$100,000 + \$40,000 + (0.25 \times \text{payroll})$. Solving for payroll in this equation indicates that the break-even payroll is $240,000 per year. Therefore, if the company's total labor payroll is less than $240,000 annually, the company did not break even by hiring the inspector. On the other hand, every $100 of payroll over $240,000 yearly means that the hire was even more beneficial.

Benefit-Cost Ratio

Another common analytical method is the *benefit-cost* ratio (B/C), the method, which has historically been required for all large projects involving federal funds. The best aspect of this method is its simplicity:

dividing the sum of all of the benefits associated with an investment by the sum of all of its costs. Projects with a benefit-cost ratio greater than 1 are typically acceptable. But this simple application of BCR analysis shares a common critical flaw with the payback period: the time value of money is ignored. Consider the two projects in Figure 6. Both projects have a benefit-cost ratio of 1.0, but Project A is clearly more attractive than Project B because the benefits accrue shortly after the outlays. If the cash inflows from project A were put into a savings account as soon as they were received, one would end up with a higher amount at year 5 than if project B had been chosen.

The flaw in the simplest benefit-cost analysis technique discussed above can be resolved by including discounted cash flows—in other words, by including benefits and costs that have been multiplied by appropriate factors of present–future worth.

Internal Rate of Return

The analysis of the *internal rate of return* (IRR) is a powerful, precise tool for evaluating candidate investments. The process identifies the timing and amounts of all cash flows, then uses a spreadsheet, a programmable calculator, or specialized software to calculate the interest rate that would make the outgoing cash flows equivalent to the incoming cash flows. Many companies establish a minimum acceptable rate of return and reject any projects that have a projected IRR lower than their minimum acceptable rate of return.

The main advantage of the IRR method over the payback-period and benefit-cost-ratio methods is that it reflects the time value of money. Another strength of the IRR method is that it provides an objective way to compare one acceptable project against another. For example, all other things being equal, a project with an IRR of 10 percent is clearly superior to a project with an IRR of 8 percent.

The main disadvantage of the IRR method is that it is somewhat difficult to explain to individuals lacking backgrounds in financial analysis, and it is very difficult to calculate without a specialized computational tool such as the IRR function in *Excel*. A lesser-known weakness of IRR is that a series of cash flows may have more than one IRR (although this is rarely a problem, because few cash flows have U-shaped patterns).

Table 3 shows the contents of a spreadsheet used to calculate the IRR of the cash flow diagrams in Figure 6. The contents of the IRR cell have been pasted to show that the IRR function has been applied to cells B6 through B11 and E6 through E11, resulting in IRR values of 0 and 14 percent, respectively.

TABLE 3

Spreadsheet Internal Rate of Return Analysis of Figure 6

Year	Cash Flow	Year	Cash Flow
0	−5000	0	−5000
1	1000	1	1000
2	1000	2	1000
3	1000	3	1000
4	1000	4	1000
5	1000	5	1000
IRR	0% = IRR (B6:B11)		14% = IRR (E6:E11)

Net Present Worth

The financial analysis method most widely accepted today is *net present worth*, which is sometimes referred to simply as *present value*. It involves summing the discounted cash flows—that is, adding the product of each future cash flow and the appropriate discount factor. Outgoing cash flows are negative, and incoming cash flows are positive. Projects with net present worth greater than 0 are considered worthwhile.

One advantage of this method over the payback-period and benefit-cost-ratio methods is that it reflects the time value of money. Indeed, the choice of the exact value of money over time (the discount rate or interest rate) is the most important decision that must be made in the analysis. The higher the discount rate chosen, the higher the discount factors associated with each cash flow. (Note how the discount factors in Table 1 decrease as the interest rate increases.) In other words, the higher the assumed interest rate, the less future cash flows affect the net present worth. The discount factor for a single payment five years in the future is 0.822 if a 4 percent interest rate is assumed and 0.621 if a 10 percent interest rate is assumed.

Given that the discount rate chosen has such a significant effect on the results of a net present worth analysis, it is important to choose wisely. As stated earlier, most companies establish a minimum rate of return and dictate that this number be used as the discount rate in all financial analysis. Some companies choose a discount rate that reflects the interest rate at which they can borrow money. (If a project cannot even generate enough cash inflows to cover the cost of borrowing the money used to fund it, it is certainly not worthwhile.) Most companies establish a discount rate equal to what they see as their true opportunity costs, or the highest rate of return that the company could obtain by investing it in other projects, or even in other companies. Sophisticated financial analysts use a method called the capital assets pricing model (CAPM) to establish their discount rate. CAPM is powerful and logical, reflecting the special risk that a project represents relative to the risk inherent in the company's existing portfolio; but it is more complicated than is necessary for most safety-investment decisions.

Annual Cost Analysis

Like present worth analysis, this method uses discount factors to adjust each future cash flow based on the assumed time value of money. Unlike NPW, which brings all of the cash flows back to year 0, this method calculates the equivalent net costs or benefits on an annual basis. It is a preferred method for managers, who evaluate projects by asking, "How much will this cost me (or benefit me) each year?"

Rather than using the discount factors that apply to present or future worth, this method uses discount factors that apply to uniform series. Another name for this analysis is the equivalent annual benefit and cost analysis, because all cash flows are converted to their equivalent in terms of an annually recurring amount. The formula for calculating the discount factor associated with converting a future cash flow (F) to an equivalent annual amount is given in Equation 5. Note that this equation assumes the annual cash flows begin in year 1 and continue through year n. As such, there will be two cash flows in year n, A and F. Most engineering economic textbooks include these factors in their compound interest tables.

$$(A/F, i, n) = \frac{1}{(1 + i)^n - 1} \quad (5)$$

As an example, compare the Equivalent Annual Costs of the two cash flow diagrams seen in Figure 5. $EUAC_A = (-12{,}000 \times 0.2034) + 2000 = -440.80$. $EUAC_B = (-12{,}000 \times 0.2034) + (6000 \times 0.1774) + (6000 \times 0.1434) = -516.00$. Because project A has a lower negative annual cost, it is more attractive than project B.

INTRODUCTION TO BUDGETING

The chapter thus far has discussed fundamental economic analysis assumptions and tools for analyzing whether to pursue specific safety investments. Once the decision has been made to pursue a project, the budgeting process should occur. A budget is a financial plan that establishes specific amounts of cash (and sometimes employee hours) that are expected to be spent on specific activities. Budgeting accomplishes a number of related purposes discussed very briefly below.

Budgeting is a form of *prioritization*. By establishing a budget for a safety-related expenditure, one is securing management approval for this expenditure and is decreasing the chance that unexpected cash flow problems, or a manager's bad day, will prevent a planned safety expenditure from taking place.

Budgeting is an important part of *cost control*. Expenditures should be tracked and compared against expected expenditures for each point in time. Significant differences between actual and expected expenditures indicate either poor tracking (in that actual expenditures are not being assigned to the correct budget item), unexpected circumstances, or poor initial estimation and budgeting. This indicates that budgeting is part of the Plan–Do–Check–Act (PDCA) cycle of process improvement. Budgets are part of the short-term operational *plan*. This plan is then executed as the *do* step. During operational execution, management should *check* by comparing actual expenditures to planned expenditures. If a significant difference exists between the two, managers *act*, resolving any unforeseen problems and, if necessary, adjusting future budgets to prevent future deviations from plans.

Budgeting is an important part of decision making. A manager must estimate the costs to include in an economic analysis. This estimate usually forms the basis for one's budget. But decision making does not end once the budget is established. If actual expenditures significantly exceed planned expenditures at any time, it may be appropriate to reevaluate the entire project. Many managers forget that *sunk costs* should be disregarded in decision making. They subconsciously think that if they have invested money in a project, the project must be pursued at all costs, feeling that abandoning the project indicates that their initial decision to pursue the project was flawed. In some cases, the initial decision *was* flawed—and continuing to pursue the project will only compound the problem. Heed the maxim and "don't throw good money after bad." Managers should ignore what has been spent in the past (sunk costs) and only continue to pursue a project if the revised estimate of *future* costs is outweighed by revised estimates of future benefits.

The Budgeting Process

Safety engineering expenditures are typically associated with one-time projects, not with recurring expenditures. Nevertheless, project expenses must always be included in a company's annual budget. Company budgets are typically established for the company's fiscal year, which may end on December 31, June 30, or September 30. The process of establishing the company budget typically is begun approximately midway through a fiscal year and is completed several months before its end.

The overall expenditure associated with a safety engineering project is typically broken down into a number of specific items to facilitate cost control and project management. For example, a project may be broken down into discrete tasks, each with an obvious beginning and end. The budget for each task is often further broken down into labor, materials, equipment, and, sometimes, overhead. Actual expenses for similar items on past projects may be used to help establish the budget for each item. Because budgets are often managed by sophisticated cost-accounting software, each cost item is typically assigned a unique cost code or number used both for entering and reporting purposes. Such account codes eliminate the need for entering awkwardly long word descriptions. In construction and other industries, this numbering system is often referred to as the *work breakdown system*.

Depending on the industry, the company, and the cost-accounting software used, overhead costs associated with a task may be budgeted and tracked as individual cost items or as part of an overall overhead mark up. For example, the salary of a safety project manager may be budgeted as a specific cost item or simply included in an overall overhead percentage that is applied to the budgeted direct costs.

It was mentioned earlier that an important step in cost control is the comparison of actual costs with expected costs at various points during a project's execution. It should be noted that it is often appropriate to identify the pattern over time associated with each cost item. Many project costs are not incurred over time in linear fashion. For example, project labor costs typically exhibit an S shape as costs are initially incurred slowly, then rapidly increase as the crew becomes productive, and then decrease as the project nears completion. Other project costs may occur at discrete intervals, such as the ends of fiscal-year quarters. Managers who assume that cost items will be incurred uniformly over a time period in cases when actual costs occur nonlinearly may make unnecessary and inappropriate cost-control decisions.

The Dark Side of Budgeting

Although budgeting is an important administrative process in all organizations, organizational budgeting also has its "dark side." How future cash flows almost always hold some degree of uncertainty, as discussed earlier, means there is always a risk that the best estimates of future costs may be low. Most managers understandably want to reduce that risk by proposing a budget that includes a contingency amount; but two problems arise when including such buffers in budgets. One problem is that allocating contingency resources to a budget prevents those resources from being used for other good projects. A second and even worse problem is that many managers will attempt to hide their overbudgeting by inefficiently spending any remaining funds as the end of the fiscal year approaches.

It should be obvious from this chapter that decisions about budget allocation should be based on which projects will provide the greatest return on investment to the company as a whole. Another dark side of budgeting is that some managers propose excessively large budgets, because the larger the budgets they manage, the more status, power, security, and salary they personally can enjoy. Because managers enjoying power and status typically are loathe to brook any reduction, it is not uncommon for managers to submit budgets for the next year that are based mostly on similar budgets in years past. In recent decades, a process called zero-based budgeting has emerged to combat such budget inertia. Instead of assuming that each project or manager will receive approximately the budget they had previously, managers and projects are assumed to have no budget until they justify in detail why they should receive any funds at all.

Conclusion

Successful safety professionals should be able to identify which safety-related initiatives represent good investments for their organizations. This chapter has provided fundamental financial analysis concepts—such as the time value of money, direct and indirect costs and benefits, and the uncertainty of future cash flows—that safety managers should integrate into their decisions and their communications with superiors. This chapter has summarized five objective methods of economic analysis commonly used to evaluate investments: payback period, benefit-cost ratio, rate of return, net present worth, and annual cost analyses. Payback period and benefit-cost-ratio analyses were identified as inferior methods because they fail to consider the time value of money. Sensitivity analysis was identified as an important step in the decision process because it requires a methodical consideration of the possible variations in cash flows and of whether such variations dramatically change the outcome of the analysis.

Acknowledgments

The help of Dr. Tim Bushnell at NIOSH in identifying relevant literature, as well as comments of three anonymous reviewers, were much appreciated for the first edition.

References

Adams, S. "Financial Management Concepts: Making the Bottom-Line Case for Safety." *Professional Safety* (August 2002), pp. 23–26.

American Society of Safety Engineers (ASSE). 2002. *The Return on Investment for Safety, Health, and Environmental (SH&E) Management Programs* (accessed June 11, 2010). www.asse.org/practicespecialties/bosc/bosc_article_6.php

Behm, M., A. Veltri, and I. Kleinsorge. "The Cost of Safety." *Professional Safety* (April 2004), pp. 22–29.

Bird, F. E. 1996. *Safety and the Bottom Line*. Loganville, GA: Institute Publishing.

Blank, L., and A. Tarquin. 2005. *Engineering Economy*. 6th ed. New York: McGraw-Hill.

Brady, W., J. Bass, R. Moser, Jr., G. W. Anstadt, R. R. Loeppke, and R. Leopold. 1997. "Defining Total Corporate Health and Safety Costs—Significance and Impact." *Journal of Occupational and Environmental Medicine* 39(3):224–231.

Brandt, J. 1999. "Hitting the Injury Iceberg." *Ergonomics Supplement*, pp. 160–165.

Brody, B., Y. Letorneau, and A. Poirer. 1990. "An Indirect Cost Theory of Work Accident Prevention." *Journal of Occupational Accidents* 13(4):255–270.

Hinze, J. 2000. "Incurring the Costs of Injuries Versus Investing in Safety." In *Construction Safety and Health Management*. Edited by R. J. Coble et al. New York: Prentice-Hall.

Kinn, S., S. A. Khuder, M. S. Besesi, and S. Woolley. 2000. "Evaluation of Safety Orientation and Training Programs for Reducing Injuries in the Plumbing and Pipefitting Industries." *Journal of Occupational and Environmental Medicine* 42:1142–1147.

Labelle, J. A. "What do Accidents Truly Cost?" *Professional Safety* (April 2002), pp. 38–42.

Lanoie, P., and S. Tavenas. 1996. "Costs and Benefits of Preventing Workplace Accidents: The Case of Participatory Ergonomics." *Safety Science* 24(3):181–196.

Leigh, J. P., S. Markowitz, M. Fahs, and P. Landrigan. 2000. *Costs of Occupational Injuries and Illnesses*. University of Michigan Press.

Linhard, J. B. 2005. "Understanding the return on health, safety and environmental investments." *Journal of Safety Research*, ECON proceedings 36:257–260.

National Institute for Occupational Safety and Health. 2004. *Does It Really Work? How to Evaluate Safety and Health Changes in the Workplace* (last accessed June 13, 2010). U.S. Department of Health and Human Services (DHHS), NIOSH Pub. No. 2004-135. Available for download at www.cdc.gov/niosh/docs/2004-135/

Newnan, D. G., T. G. Eschenbach, and J. P. Lavelle. 2004. *Engineering Economic Analysis*, 9th ed. New York: Oxford University Press USA.

Occupational Safety and Health Administration. 2010. *Making the Business Case for Safety and Health* (accessed June 11, 2010). www.osha.gov/dcsp/products/topics/businesscase/index.html

———. 2010. OSHA's "$AFETY PAYS" Program (accessed June 11, 2010). www.osha.gov/dcsp/smallbusiness/safetypays/index.html

———. 2010. "Safety & Health Management Systems eTool" (accessed June 11, 2010). www.osha.gov/SLTC/etools/safetyhealth/mod1_costs.html

———. 2010. *Safety Success Stories* (accessed June 11, 2010). www.osha.gov/dcsp/compliance_assistance/success_stories.html

Occupational Safety and Health Administration, Abbott Laboratories and Georgetown University Center for Business and Public Policy. 2005. "The Business Case for Safety: Adding Value and Competitive Advantage" (powerpoint file) (accessed June 12, 2010). www.osha.gov/dcsp/success_stories/compliance_assistance/abbott/abbott_casestudies/index.html

Oxenbaugh, M., P. Marlow, and A. Oxenbaugh. 2004. *Increasing Productivity and Profit through Health & Safety*. London: CRC Press.

Veltri, A., and J. Ramsay. 2009. "Economic Analysis: Make the Business Case for SH&E." *Professional Safety* 54(9):22–30.

Veltri, A, M. Pagell, M. Behm, and A. Das. 2007. "A Data-Based Evaluation of the Relationship between Occupational Safety and Operating Performance." *Professional Safety* 49(1):1–21.

Benchmarking and Performance Criteria

Christopher Janicak

6

LEARNING OBJECTIVES

- Develop various leading, trailing, and current indicators that can be used to assess a safety program activity.

- Develop an effective safety performance measurement program designed to address a safety issue in the workplace.

- Evaluate the performance of a safety program activity using a variety of measurement techniques.

- Incorporate a continual improvement process into safety performance activities in the workplace.

A QUESTION OFTEN POSED by a safety manager is "Are my safety activities working?" Safety performance should be evaluated in an organization in the same manner as productivity and other aspects of the business. Areas to evaluate include determining the overall effectiveness of a particular intervention, where a company's safety performance stands with regard to other companies, identifying potential impediments to safety success, and determining the trends in accidents and losses over time.

An integral part of any safety activity should include techniques for developing goals and objectives, collecting data to measure the success of the safety interventions, evaluating the results, and implementing the appropriate corrective action. The manner in which safety performance is measured can range from developing safety performance measures unique to a particular organization to using existing performance measures and standards to benchmarks derived from similar industries and organizations.

Regardless of the approach, the measurement and evaluation of safety performance requires a carefully structured program of planning, establishing goals and objectives, identifying valid measures, conducting proper data analysis, and implementing appropriate follow-up measures.

INTERVENTION EFFECTIVENESS

The overall effectiveness of the safety program in reducing accidents, controlling losses, and improving the overall working conditions in an organization is contingent upon a number of aspects, all of which are interrelated. These aspects can be broken down into seven main categories (Swartz 2000, 42):

1. Management commitment and support
2. Employee participation

3. Control or elimination of hazards
4. Integration of safety and health throughout the organization
5. Job safety analyses (JSAs)
6. Employees who are selected and trained for their positions
7. Safety and health professionals who are up to date on scientific, technical and regulatory, and legislative knowledge.

Incorporating these strategies into a safety intervention program can increase overall effectiveness and lead to measurable outcomes. Management commitment and support includes having adequate financial resources and staffing for safety. To gain acceptance of safety activities and to provide input into solutions, employee participation is vital for any successful safety intervention. Employee participation in a safety metrics program can involve defining safety performance measures, collecting data, and developing countermeasures to help the organization improve.. The key to the success of a safety intervention is the organization's ability to identify and correct hazards through the use of a well-structured hazard recognition, evaluation, and elimination program. Finally, a safety program cannot function adequately if it is disassociated from the rest of the activities in the organization. Safety should be integrated into all aspects of the company's activities, assigning responsibilities for meeting the safety goals and objectives to all employees.

Along with the identification and control of physical hazards in the workplace, the identification and control of unsafe job procedures is equally important. Job hazard analyses (JHAs) involve the identification and elimination of hazards associated with the job tasks in the workplace. JHAs are also useful in developing safe job procedures used in training new employees. Performing safe job procedures requires training across all aspects of safety in the workplace, from the safety procedures for performing a job task to the procedures necessary to implement various interventions, such as hazard recognition programs and emergency response procedures.

Finally, managing safety in the workplace requires specialized managerial and technical skills. The overall administration of safety requires someone who is knowledgeable in the technical, managerial, and legal aspects of safety.

SAFETY MANAGEMENT SYSTEM

The effectiveness of any safety intervention can be tied to two main aspects of the overall safety program: the existence of a safety management system and an organizational culture that is supportive of the safety efforts. The Occupational Safety and Health Administration (OSHA) defines a safety management system as being comprised of four areas, all of which are necessary for a safety and health program to be effective in meeting its goals and objectives. The components of the safety and health management system include management leadership, employee involvement, work-site analysis, hazard prevention and control activities, and safety and health training (OSHA 2004a).

Management Leadership and Employee Involvement

Without management leadership for safety, a safety program can be almost guaranteed to be ineffective. Through their actions, members of senior management display the importance that safety plays in an organization. Including safety performance as part of the overall organizational goals is one way management conveys this importance. If safety is not perceived by the employees to be important to management, then it will almost certainly not be seen as being important by the workers. Where management has placed safety on a par with other functions, they must be genuinely committed to following through or employees will not abide by company policies (Swartz 2000, 42). Getting employees involved in the development and implementation of safety program tasks increases the chances that their programs will be accepted and followed by the employees.

Work-Site Analysis

Work-site analysis involves the identification of hazards with the goal of correcting hazardous conditions before an accident occurs. Tools to consider as part of the work-site analysis include conducting property hazard assessments, environmental audits, accident investigations, and job hazard analyses, and analyzing accident data. Proactive safety programs are implemented with the goal of preventing potential accidents and the losses from those accidents before they occur. Reactive safety programs, on the other hand, focus their attention on activities aimed specifically at the causal factors attributed to accidents and losses that have already occurred.

Hazard Prevention and Control

Hazard prevention and control includes those program components designed to prevent accidents from occurring, and the components intended to minimize their severity should an accident occur. Examples of programs aimed at hazard prevention and control include preventive maintenance programs and emergency preparedness. A recognized hierarchy for hazard control is elimination, substitution, engineering, warning, administrative action, and the use of personal protective equipment (PPE) (ANSI 2005).

Safety and Health Training

The fourth component of a safety management system is safety and health training. The training should ensure that employees at all levels of the organization are aware of safety and health policies and procedures that may impact them. Additionally, task-specific safety and health training should be provided to employees with unique exposures to hazards on the job.

To evaluate the safety and health management system, OSHA has developed, as part of their outreach programs, an evaluation tool referred to as the Safety and Health Program Assessment Worksheet (OSHA Form 33). As part of the assessment, consultants review an employer's existing safety and health management program to identify elements considered adequate and elements that need development or improvement. To assist employers in meeting their training obligations, OSHA published the training requirements in *OSHA Standards and Training Guidelines* (OSHA 1998). This document provides employers with guidance on how to identify training needs, develop a training program, and evaluate the effectiveness of the program.

Components of a Comprehensive Safety and Health Program

The components identified as necessary for a comprehensive safety and health program include the following (OSHA 2008b):

1. Hazard anticipation and detection programs, including hazard surveys, self-inspections, and accident investigations
2. Hazard prevention and control measures, including the use of engineering controls, personal protective equipment, emergency response plans, and adequate medical care for employees
3. Planning and evaluation programs, including data collection and analysis methods, development of safety goals and objectives, and a review of the overall safety and health management system
4. Administration and supervision activities, including coordination of safety and health program activities, accountability mechanisms, and safety responsibilities communicated to those who must perform the duties
5. Safety and health training, encompassing new employee orientations, supervisor safety training, and management safety training
6. Management leadership and commitment is vital to the success of any safety program. Performance measures include adequate resource allocation for safety and top management involvement in the planning and evaluation of safety performance.
7. Effective safety performance requires employee participation in all areas of the planning, evaluation, and implementation of safety program tasks. Employee involvement can take on many forms, including employees involved in the decision-making process for safety, and participation in the detection and control of hazards.

Occupational Safety and Health Management System Cycle

The *American National Standard for Occupational Health and Safety Management Systems* (ANSI/AIHA A10) defines an occupational health and safety management-system cycle as an initial planning process and the implementation of the management system, followed by a process for checking the performance of these activities and taking appropriate corrective actions (AIHA 2005). This is then followed by a management review of the system for suitability, adequacy, and effectiveness against its policy and the ANSI standard (AIHA 2005).

Components of this cycle include the plan-do-check-act cycle along with management leadership and employee participation, management planning activities, implementation, checking and corrective action, and management review (AIHA 2005).

The purpose of this cycle is to ensure that continuous improvement activities are systematically incorporated into the organization's management functions, resulting in a coordinated effort to continually improve safety performance.

SAFETY CULTURE

An organization's culture consists of its values, beliefs, legends, rituals, mission, goals, performance measures, and sense of responsibility to its employees, customers, and community, all of which are translated into a system of expected behavior (Swartz 2000, 18). The safety culture of an organization defines how the organization values and perceives safety in the workplace. This safety culture plays an important role in determining the success of safety and health activities. If management promotes a culture in which safety is perceived as not being important to the organization, then the employees will perceive safety as something that is not important. It is the organization's culture that determines whether the safety program as a whole will be effective. An assessment of the safety culture should include asking questions such as the following (Weinstein 1997, 24):

- Is there a strong safety culture established with no tolerance for unsafe practices?
- Is the cultural goal zero injuries?
- Are health and safety procedures followed all the time?
- Is there a vision of a safe work environment, and do all employees share in it?
- Do employees value safe behavior, themselves, and their continued well-being?
- Is the management style and culture nonautocratic with a win-win atmosphere?
- Is there a trusting relationship between management and employees?
- Do employees believe that safety is a company priority?

In organizations with a strong safety culture, the following characteristics exist (Weinstein 1997, 24):

- Executives and managers visibly support safety with no contradictory decisions, and they accept full accountability.
- Employees are involved with safety and their views are sought and acted upon.
- Supervisors' actions support safety, including recognizing and appreciating safe work practices and behaviors.

It is accepted in the safety profession that there is a relationship between an organization's culture and safety performance and that the organization's culture can be measured and managed (Swartz 2000, 82–83; Mohamed 2000, 384). Methods used to measure the safety culture in an organization include employee surveys directed at their perceptions about management leadership for safety, reinforcement by management to report hazards, employee attitudes and perceptions about safety, how employees view the management and supervision of safety, and whether they feel there is a real and genuine commitment for safety.

MEASURING SAFETY EFFECTIVENESS

In occupational safety and health, the need for a particular intervention can be determined by legislation in which a regulation stipulates that a particular safety and health activity be provided in addition to other areas not regulated by standards, such as ergonomics. It can also be determined by analysis and investigation. For example, an analysis of the work site and loss data may indicate the need to prevent back injuries. Once the intervention is in place, many times a more difficult question presents itself: "Is the safety intervention working?"

The methods safety professionals use to answer this question vary widely. Some companies count the number of people injured at the end of the year, and others may use a continuous improvement process. As with any intervention in the workplace, an organization must determine if the activities implemented are effective in meeting the organization's goals and objectives. Safety activities are no different than any other business activities. Over the years, it has become more commonplace for the safety professional to tie safety activities to results in an effort to show how improving safety activities equates to improving business operations.

Historically, the effectiveness of a safety activity has been measured in terms of the number of accidents incurred, the organization's OSHA recordable incidence rates, the dollars spent on accidents, and the costs for insurance coverage. There is no one way to measure safety and health program effectiveness; rather, a systems approach is necessary (Swartz 2000, 98). These multiple methods for measuring safety performance include an approach in which leading, trailing, and current indicators are used.

As methods for continual improvement evolved in the workplace along with statistical process control, the safety profession has slowly moved toward some of these methods now routinely used in other aspects of the organization's management structure. Safety managers are increasingly held accountable for their activities and must show management how their activities positively impact the organization.

With an ever-increasingly global economy, and international standards becoming the framework by which management practices are designed and monitored, safety practices have evolved to systems approaches for continual improvement. An organization must accurately and validly assess where they are in terms of their safety performance, how they decide where they would like to be, and what needs to be done to get there (Petersen 1996, 3).

In recent years, much research has been conducted to evaluate the effectiveness of interventions designed to improve safety performance (Al-Mutairi and Haight 2009, Iyer et al. 2004). Through modeling techniques and statistical analysis, it is possible to optimize the effects of the the safety and health interventions by decreasing injury rates and property damage with less costly programs.

Valid Measurements

Measuring safety performance is a critical step in the safety performance improvement process. The purpose of safety performance measurement is to determine if the goals and objectives have been met. The measures selected to monitor and evaluate safety performance must be valid and reliable. Valid performance measures are measures that are true indicators of performance. There must be a relationship between what is being measured and safety performance.

Because follow-up action is planned and implemented based on the outcomes of the performance measures, it is only logical that the corrective actions are also valid means for improving performance. For example, a safety manager determined that an indicator of the number of cumulative trauma disorder (CTD) injuries reported was the number of employees that successfully completed CTD injury prevention training. Using this measure, the safety manager tracked the number of employees trained each month and the number of CTD injuries reported. The safety manager found that as the number of employees in the facility trained on CTD injuries increased, so did the number of CTDs reported, indicating that the training was unsuccessful in reducing the number of injuries.

What the safety manager failed to take into account was the fact that the training also included early symptom reporting procedures and information about the early symptoms of CTD injuries. Thus, using the completion of CTD training as an indicator of CTD injury prevention may not be considered a valid measure because the training introduced a confounding factor—the early reporting of CTD symptoms. Variables confounding in data research are variables whose individual effects upon an outcome cannot be readily measured. In some cases, statistical procedures may be used to control for this confounding.

Another important trait of any measure used to evaluate safety performance is *reliability*. The reliability of a performance measure is the consistency of results obtained through the measurement. This consistency means that the same results are obtained when the measurement is taken multiple times. A measurement used to describe the number of CTD injuries suffered must be well-defined to ensure the reliability of the data collected. An unreliable measure can yield different numbers when measured by different people. Data must first be proven to be reliable before it can be evaluated for validity. Otherwise stated, unreliable data is always considered invalid. Reliable data may or may not be valid.

Reliability of data can be statistically evaluated using a variety of techniques. Two examples of these methods include the test-retest method and the split-half method. In the test-retest method, a performance indicator is measured multiple times. If the measurements are highly correlated, meaning the same results are obtained over the multiple trials, the measurement technique and data can be shown to be reliable.

The split-half method is commonly used with tests and survey instruments. With the split-half method, the items are randomly distributed throughout the instrument. If the items are consistently measuring the same outcome, one would expect to find a strong correlation when comparing the first half of the responses to the second half.

Leading Indicators, Trailing Indicators, and Current Indicators

The effectiveness of a safety activity should be measured via three indicators: leading indicators, current indicators, and trailing indicators. Trailing indicators are the most common measures used by safety professionals. *Trailing indicators* are those measures that indicate the results of an intervention strategy after the fact. Examples of trailing indicators include lost-workday rates, the number of injuries over a period of time, and the losses incurred by the organization. Some reasons why trailing indicators are so widely used to measure safety performance include the availability of data to make such measurements, the influence of OSHA's record-keeping guidelines, and use of various OSHA rates and measures of safety performance in the United States. One major downside of using trailing indicators is that they are measuring unwanted events after the fact, thus providing no means for implementing improvement strategies to impact their outcomes.

Current indicators measure the current status of the organization's safety performance. An example of a current indicator is the number of safety audits conducted up to a particular point in time. A positive outcome from using current indicators is that as soon as the measure is obtained, action can be taken immediately to improve the measure and thus improve safety performance.

Leading indicators are those measures that are correlated to future safety performance. For example, participation in safety training may be an indicator as to whether employees suffer back injuries on the job. Measuring the number of workers trained at a point in time may be indicative of the number of back injuries expected in the future. As with current indicators, leading indicators provide the safety manager with information that can be acted upon today with positive results on the safety performance in the future. A key to using current and leading indicators is that these measures must be directly correlated to safety performance. Without this relationship, a safety manager may find that activities taken to improve safety performance based on uncorrelated measures will have no effect on safety performance.

Safety performance should *not* be measured using only one or two performance measures. Instead, it should be measured with a variety of leading, current, and trailing indicators that have been shown to be correlated to safety performance in the workplace. When selecting these performance measures, keep in mind that the data needed for the performance measure should be valid, reliable, and readily available.

When using multiple measures, the data's main effects and interactive effects become important when interpreting the results. Main effects are the variables examined separately in order to determine their role in influencing the outcome measure. For example, a safety manager wishes to determine the influence the age of the worker and the number of training sessions attended have on the number of injuries reported over a given period. The age of the worker and the number of training sessions attended can be considered the main effects. Next, the safety manager wishes to determine the influence that both age and number of training sessions attended together have on the number of injuries reported. When examining the two variables simultaneously, the safety manager is assessing the interactive effects of the two variables.

BENCHMARKING

Benchmarking, measurement, and evaluation are all essential for program success (Lack 2002, 684). The benchmarking process establishes a standard that the company has determined signifies successful performance. *Benchmarking* is a technique for measuring an organization's products, services, and operations against those of its competitors, resulting in a search for best practices that will lead to superior performance (Hoyle 2003, 15). Benchmarking safety performance entails identifying similar organizations with outstanding safety performance and identifying the key aspects of their activities that make them stand out.

Benchmarking is more than taking another organization's safety programs and copying them. Much research is necessary to be able to identify those aspects of safety activities that result in superior performance, and much work is required to tailor them so similar outcomes can be duplicated in another organization. Meaningful benchmarks are typically set using successful performance results from similar industries and other facilities. The benchmarking process can be completed in six steps (Pierce 1995, 177–178): survey programs, identify solutions, prioritize, develop a plan, implement the plan, and then follow up.

Surveying

The first part of benchmarking is surveying frontrunning programs or organizations. This step is the most crucial in the entire process. Identifying who the best organizations are and what they are doing to generate exemplary safety performance is critical in establishing benchmarks and program priorities.

Identifying Solutions

The second part of benchmarking is identifying the complementary solutions used by the target organization or program. As stated previously, the benchmarking process is not merely copying what other successful organizations are doing, but incorporating their programs into your organization in a manner that fits the organization structure and goals.

Prioritizing

Part three of benchmarking is prioritizing growth opportunities from the list of complementary solutions. The purpose of prioritization is to determine which program changes will provide the organization with the largest improvement in business and safety performance.

Planning

The next part of the process involves developing a plan to achieve the goals. Incorporating changes in an organization will take time and careful planning. The programs identified as being crucial for success must be tailored to the organization.

Implementing

Implement the plan. Adequate personnel and resources must be made available to ensure the benchmarking plan is carried out. Inadequate resources in the implementation phase, a lack of commitment, and a lack of motivation to continue implementing the benchmarking plan will result in poor results.

Following Up

Benchmarking is a dynamic process. Follow-up activities include monitoring to ensure the changes are meeting the needs of the organization. Just because they were found to be successful in one organization

CASE STUDY

Benchmarking a Safety Measure

An organization was experiencing ever-increasing workers' compensation costs due to employee back injuries. To control these costs, the safety director decided to apply a benchmarking approach. First, companies that had been recognized by the industry as leaders in safety were identified and invited to participate in benchmarking focus groups. In these focus group meetings, the activities that were being used to control workers' compensation costs were identified and prioritized in terms of their effectiveness. Following the focus group meetings, the safety director developed a plan to tailor the activities to best meet the needs of his facility and implement them. Using a continuous improvement approach, results from the cost-control activities were measured and further interventions implemented based on the measurable results.

does not necessarily mean they will achieve the same results in another. Follow-up may involve modifying the activities or identifying new ones to achieve the desired safety performance.

QUALITY CONTROL

Quality control is a universal management process for conducting operations in order to provide stability, for preventing adverse change, and for maintaining the status quo (Juran and Godfrey 1999, 4.2). *Process control* is about maintaining variation in a process at a level where the only variation present is random and the process is stable and, therefore, predictable (Hoyle 2003, 28). The major distinction between the two is that quality control focuses on outputs and process control focuses on inputs.

Continuous improvement is about improving the efficiency and effectiveness of products, processes, and systems that are under control (Hoyle 2003, 28). Continuous improvement efforts are directed toward both inputs and outputs.

Safety performance, like other aspects of business, can be managed using the tools and techniques found in quality control. Quality safety performance begins with proper planning and the development of performance goals and objectives. Methods for measuring this performance, commonly referred to as safety metrics, are then developed to define measures that are indicative of acceptable performance. Data are collected and analyzed, making comparisons against the established levels of acceptable performance. When gaps are identified between acceptable performance levels and actual performance levels, action is warranted. This quality control process is known as a *continuous improvement process*.

Continuous Improvement

During the 1990s the use of continuous improvement processes increased dramatically in the business world. In the context of ISO 9000, there is no difference between continuous improvement and continual improvement. Improvement that is continuous has no periods of stability; it is attainable all the time. Although the rate of change may vary, improvement does not stop. In reality, there are periods of stability between periods of change, and therefore, continual improvement is a better term to describe the phenomenon (Hoyle 2003, 29).

Continuous improvement is the process of establishing performance measures with a desired goal, implementing an intervention designed to meet that goal, measuring the performance, and implementing change in the intervention until the desired goal is met.

Juran and Deming say that putting out fires is not improvement of the process, and neither is the discovery and removal of a special cause detected by a point out of control (Hoyle 2003, 28). This only puts the process back to where it should have been in the first place. If there is no status quo (no normal level), action needs to be taken to establish a normal level, that is, bring operations under control. When a process is in control, data measurements are consistent without wild fluctuations and wide ranges. You can only improve what is already under control. Bringing operations under control is not improvement.

Plan-Do-Check-Act

The Plan-Do-Check-Act (PDCA) cycle has been adopted by a variety of industries in their quest for safety performance improvement. This cycle has also been incorporated into standards such as ISO 9001:2008, the ISO 14000 family of standards, and ANSI/AIHA A10:

2005. Various people in the United States and Japan have been associated with the early evolution of the PDCA cycle, including Deming, Shewhart, and Mizuno (Juran and Godfrey 1999, 11.16). This continuous improvement approach has been the cornerstone of a variety of approaches to both safety and process control.

First, safety activities and performance goals are defined and prioritized in the Plan phase. Next, the safety activities are implemented in the Do phase, followed by measurement of the results of the activities and comparisons of the results with the planned or desired outcomes in the Check phase. In the Act phase, if the desired performance levels are not achieved, then changes in the activities may be warranted to achieve the desired outcomes. If the performance goals are successfully met, then modifications to the planned outcomes can be made so that further improvements in performance can be planned. By repeating this cycle and planning for better safety performance with each successive time through the process, continuous improvement can be successfully planned, implemented, measured, and achieved.

The first edition of the ISO 9000 standards sees PDCA management as compatible with the contemporary concept that all work is accomplished by a process (Juran and Godfrey 1999, 11.16). Safety management practices also lend themselves well to a PDCA process in which safety activities are planned and implemented, outcomes are evaluated, and interventions are acted upon to close the gaps between desired performance and actual performance.

STATISTICAL PROCESS CONTROL

Process control means that processes are planned, executed, and controlled such that the equipment, environment, personnel, documentation, and materials employed constantly result in meeting quality or safety requirements (Weinstein 1997, 85). *Statistical process control* involves setting quantifiable performance goals, measuring performance against those goals, and determining with a degree of statistical certainty that the performance goals have been met. If there is a difference between the desired level of performance and actual performance, then appropriate measures are taken to close the gap between the two.

An underlying assumption of statistical process control is that the data collected on performance measures is assumed to follow a normal distribution. On a control chart that is recording data compared to an average measure, for example, the center line on the chart represents the sample average, and the control limits represent the upper and lower boundaries for which any data point lying above or below the control limits have a 5 percent chance of falling there. Points falling between the control limits have a 95 percent chance of falling there. Much the same way a bell-shaped curve is interpreted, those points lying at the extremes may be considered to be significant.

Some fluctuations in performance can always be expected. This is taken into account in statistical process control. Action is taken when the performance is statistically significantly different from what is acceptable.

The key to being able to use statistical process control techniques to monitor a process is that the process must be in control. A process is considered in control when the performance measures are stable, falling within a narrow range. Large fluctuations in measures indicate that the process is not in control, and any resulting controls established for a process that is not in control may not be meaningful. With a process that is in control, performance measures can be taken and compared against established control limits. When data points fall outside of the control limits, it is an indication that some action may be necessary to bring the process back into control.

There are a number of criteria that have been developed to determine if a process is in control. Data trends that indicate an out-of-control process include one or more points outside the limits on a control chart, two or three points outside two standard deviations from the average, and four or five points outside one standard deviation from the average.

With the data indicating a process that is in control, interventions are then implemented with a goal of creating a trend in the data. For example, a safety manager wishes to decrease the number of hand injuries reported in a facility. Data is collected to determine the current extent of hand injuries reported each month, and this is used as a baseline. A control chart is created using this past data. A variety of activities designed to decrease these injuries is implemented and the performance is measured. Over the months, a downward trend in reported hand injuries is observed. The criteria for indicating a trend, either upward or downward, can include points that are lying outside of the control limit but are doing so in a pattern that is either increasing or decreasing.

When the data points consistently fall below the lower control limit when a downward trend is desired, or above the upper control limit when an upward trend in data is desired, a performance shift has occurred. It is necessary to recalculate the control chart limits using the new data. This will, in effect, further improve the performance goal and require continuous improvement.

Finally, from a practical standpoint, if statistical process control methods are to be used successfully in an organization to monitor safety performance, the data required to measure the performance must be readily accessible to the users, valid, and reliable. To aid in the data collection process, data should be designed carefully to make the data readily available.

Control Charts

Control charts are widely used in industry as the principal tools of statistical process control (SPC). In 1931, Shewhart published his classic book, *Economic Control of Quality of Manufactured Product*. The first applications of the control chart by Shewhart were on fuses, heat controls, and station apparatus at the Hawthorne Works of the Western Electric Company (Juran and Godfrey 1999, 45.3). Today, control charts can be a very useful tool for the safety professional to determine if performance is within acceptable ranges.

Control charts can be broadly classified into two distinct types. The first type is control charts in which no standard is given. These are used to determine if the data points collected vary among themselves by an amount greater than is attributed due to chance. The second type is control charts developed against a standard. These charts use a standard value based on prior experience or an established acceptable level of performance. Data are collected and comparisons are made to the acceptable standard. Control charts based on a standard are particularly used to control processes and maintain quality uniformly at the desired level (ASTM 1995, 52).

Control charts use the normal distribution, or bell-shaped, curve as the basis for their use and construction. If the safety performance measurements are truly random, then they can be assumed to follow a normal distribution. It is this assumption of normally distributed data that must be met in order to use control charts with safety performance data.

Formats of Control Charts and Their Uses

There are a variety of different types of control charts that can be developed and used to monitor safety performance. The selection of the type of control chart to be used is based on the format of the data being measured. The following guidelines should be followed when constructing control charts when a standard is given. When using a given standard, only the information regarding sample size is required in order to compute central lines and control limits. These standard values are set up before the detailed analysis of the data at hand is undertaken, and frequently before the data to be analyzed are collected. The data must exhibit not only control, but also control at the standard level and with no more than standard variability. Extending control limits obtained from a set of existing data into the future, and using these limits as a basis for purposive control of quality during production, is equivalent to adopting the values obtained from the existing data as standard (ASTM 1995, 61).

X Control Charts

The X chart (see Figure 1) is one of the most commonly used statistical process control procedures. It is used whenever there is a particular quality characteristic that one wishes to control, since the chart can use one characteristic at a time. In addition, the data must be of a measurement or variables type. Most users of process control are interested in individual

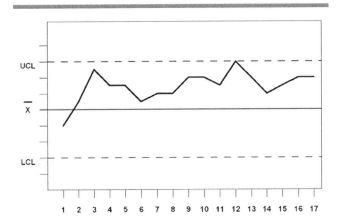

FIGURE 1. Sample X control chart

items of product and the values of a few quality characteristics on these items. Averages and ranges computed from small samples or subgroups of individual items provide very good measures of the nature of the underlying universe. They permit control and decision making about the process from which the items came. The chart for averages is used to control the mean or central tendency of the process, whereas the chart for ranges is used to control the variability. In place of the range, the sample standard deviation is sometimes used, but the range (the largest minus the smallest values in the sample) is easier to calculate and is more easily understood by the operators (Juran and Godfrey 1999, 45.5).

For averages, the central line is the established mean set for the acceptable level. The upper and lower control limits are then calculated using the formula for small samples and equal sample sizes:

Equation 1: Central Line and Control Limits for X Control Charts

Central line is a predetermined set standard: μ

$$\text{Upper Control Limit} = \mu + 3\frac{\sigma}{\sqrt{n}}$$

$$\text{Lower Control Limit} = \mu - 3\frac{\sigma}{\sqrt{n}}$$

where

μ = average number of nonconformities

n = number of cases

σ = standard deviation

CASE STUDY

Example X Control Chart

Data was collected so that ten safety audits were performed each day throughout the plant over a 10-day period. The average number of hazards identified each day was calculated. Using past data, an average of 1.1 hazards each day was to be expected with a standard deviation of 0.94. The raw data for Day 1 was as follows:

Audit Number	Number of Hazards Identified
1	0
2	1
3	0
4	2
5	2
6	0
7	1
8	2
9	0
10	2

The average number of hazards for Day 1 was calculated as follows:

$(0 + 1 + 0 + 2 + 2 + 0 + 1 + 2 + 0 + 2) \div 10 = 1.0$

The standard deviation for the sample was determined to be 0.94

The average and standard deviation are then calculated for the remaining days. The following table displays the data from which a control chart was constructed:

Day	Number of Audits Conducted	Average Number of Hazards Identified	Standard Deviation
1	10	1.0	0.94
2	10	1.5	1.1
3	10	1.2	0.89
4	10	1.1	0.93
5	10	0.98	0.99
6	10	1.3	0.96
7	10	0.99	0.99
8	10	1.0	0.93
9	10	1.2	0.95
10	10	0.93	0.98

Central line is a predetermined set standard: $\mu = 1.10$

$$\text{Upper Control Limit} = 1.10 + 3\frac{0.98}{\sqrt{10}} = 2.03$$

$$\text{Lower Control Limit} = 1.10 - 3\frac{0.98}{\sqrt{10}} = 0.17$$

The data in the table is then plotted on the control chart using the central line and control limits calculated above.

CASE STUDY

Example P Control Chart

In a p chart, the fraction of nonconforming items is determined for each sample. In this example, the samples are equal in size. A systems safety engineer collected data to determine the fraction of defective bolts making their way to the production floor from their suppliers. A random sample of 500 bolts was examined on a daily basis, and the fraction of defective bolts calculated. An acceptable p was established by the company to be 0.002. The following data table was developed:

Day	Sample Size	Number of Defective Bolts	Fraction Nonconforming (p)
1	500	1	0.002
2	500	2	0.004
3	500	1	0.002
4	500	2	0.004
5	500	4	0.008
6	500	2	0.004
7	500	1	0.002
8	500	3	0.006
9	500	4	0.008
10	500	2	0.004

Central line is a predetermined set standard: $p = 0.002$

$$\text{Upper Control Limit} = 0.002 + 3\sqrt{\frac{0.002(1-0.002)}{500}} = 0.008$$

$$\text{Lower Control Limit} = 0.002 - 3\sqrt{\frac{0.002(1-0.002)}{500}} = -0.004$$

Note: Because the lower control limit (LCL) is negative, the LCL is set to 0.

The data in the table is then plotted on the control chart using the central line and control limits calculated above.

P Control Charts

A p control chart is used to represent the fraction of nonconforming parts in a sample. Ordinarily, the control chart for p is most useful when the sample sizes are large—when n is 50 or more and when the expected number of nonconforming units (np) is 4 or more. When n is less than 25 or the expected np is less than 1, then the control chart for p may be more reliable (ASTM 1995, 64).

Equation 2: Central Line and Control Limits for P Control Charts

Central line is a predetermined set standard: p

$$\text{Upper Control Limit} = p + 3\sqrt{\frac{p(1-p)}{N}}$$

$$\text{Lower Control Limit} = p - 3\sqrt{\frac{p(1-p)}{N}}$$

where
- p = set standard proportion of cases
- N = number of cases

Np Control Charts

An np chart is used to chart the number of nonconforming parts in a sample. It is equivalent to the control charts for the fraction of nonconforming parts.

CASE STUDY

Example Np Control Chart

Using the data in the p chart example, an np chart can be developed. In this example, np represents the number of defective bolts.

Day	Sample Size (n)	Number of Defective Bolts (np)	Fraction Nonconforming (p)
1	500	1	0.002
2	500	2	0.004
3	500	1	0.002
4	500	2	0.004
5	500	4	0.008
6	500	2	0.004
7	500	1	0.002
8	500	3	0.006
9	500	4	0.008
10	500	2	0.004

Central line is a predetermined set standard: $np = 1.00$

$$\text{Upper Control Limit} = 1.00 + 3\sqrt{1.00} = 4.00$$

$$\text{Lower Control Limit} = 1.00 - 3\sqrt{1.00} = -2.00$$

Note: Because the lower control limit is negative, the LCL is set to 0.

CASE STUDY

Example *U* Control Chart

In this example, a safety manager collects data from supervisors who have conducted job observations of workers packing boxes on the production lines. The supervisors sampled 10 workers over 10 work shifts and counted the number of unsafe material-handling events. The predetermined standard was set at 0.75.

Work Shift	Number of Unsafe Events	Unsafe Events Per Worker (u)
1	4	0.4
2	6	0.6
3	3	0.3
4	6	0.6
5	8	0.8
6	6	0.6
7	5	0.5
8	4	0.4
9	3	0.3
10	3	0.3

Central line is a predetermined set standard: $u = 0.75$

$$\text{Upper Control Limit} = 0.75 + 3\sqrt{\frac{0.75}{10}} = 1.57$$

$$\text{Lower Control Limit} = 0.75 - 3\sqrt{\frac{0.75}{10}} = -0.07$$

Note: Because the lower control limit is negative, the LCL is set to 0.

particularly when all samples have the same n (ASTM 1995, 64).

Equation 3: Central Line and Control Limits for *Np* Control Charts

Central line is a predetermined set standard: np

$$\text{Upper Control Limit} = np + 3\sqrt{np}$$
$$\text{Lower Control Limit} = np - 3\sqrt{np}$$

where

p = fraction of cases
n = number of cases in the sample

U Control Charts

The control chart for nonconformities per unit (u chart) is calculated using the total number of nonconformities in all of the units divided by the number of units in the sample (ASTM 1995, 65). The u chart is desirable when inspections look at multiple characteristics of a unit. It is assumed that for each characteristic under consideration, the ratio of the expected to the possible number of nonconformities is small—less than 0.10.

Equation 4: Central Line and Control Limits for *U* Control Charts

Central line is a predetermined set standard: u

$$\text{Upper Control Limit} = u + 3\sqrt{\frac{u}{N}}$$

$$\text{Lower Control Limit} = u - 3\sqrt{\frac{u}{N}}$$

where

u = number of nonconformities per unit
N = number of cases in the sample

C Control Charts

C charts are used to chart the number of nonconformities in a sample. The c chart is equivalent to the control chart when all sample sizes have the same n. The u chart is recommended as preferable when the sample sizes vary from sample to sample (ASTM 1995, 61).

Equation 5: Central Line and Control Limits for *C* Control Charts

Central line is a predetermined set standard: c

$$\text{Upper Control Limit} = c + 3\sqrt{c}$$
$$\text{Lower Control Limit} = c - 3\sqrt{c}$$

where

c = number of nonconformities

CASE STUDY

Example C Control Chart

Using the data set in the u chart example, a c chart can be constructed. A standard of 3.0 was established as the c-chart control line by the safety manager.

Work Shift	Number of Unsafe Events (c)	Unsafe Events Per Worker (u)
1	4	0.4
2	6	0.6
3	3	0.3
4	6	0.6
5	8	0.8
6	6	0.6
7	5	0.5
8	4	0.4
9	3	0.3
10	3	0.3

Central line is a predetermined set standard: $c = 3.00$

Upper Control Limit $= 3.00 + 3\sqrt{3.00} = 8.20$

Lower Control Limit $= 3.00 - 3\sqrt{3.00} = -2.20$

Note: Because the lower control limit is negative, the LCL is set to 0.

Constructing a Control Chart When No Standard Is Given

When no standard is given as the acceptable level of performance, it must be determined using data collected from the process itself. The average and the control limits are determined using data collected from the process, and the appropriate standard is derived. The procedure for constructing all types of charts previously described when no standard is given is as follows (Juran and Godfrey 1999, 45.5):

1. Take a series of 20 to 30 samples from the process.
2. During the taking of these samples, keep accurate records of any changes in the process, such as a change in operators, machines, or materials.
3. Compute trial control limits from these data.
4. Plot the data on a chart with the trial limits to determine if any of the samples are out of control, that is, if any plotted points are outside the control limits.

If none of the plotted points are outside the trial control limits, one can say the process is in control and these limits can be used for maintaining control. If, on the other hand, some of the plotted points are outside the trial control limits, then the process is not in control. That is, there are assignable causes of variation present. In such a case, one must determine from the records in Step 2 above, if possible, the cause of each out-of-control point. Then eliminate these samples from the data and recalculate the trial control limits. If some points are outside these new limits, this step must be repeated until no points are outside the trial control limits. These final limits can then be used for future control.

Formulas for Central Lines and Control Limits for Control Charts When No Standard Is Given

Equation 6: X Charts

Control Charts for Averages (large sample more than 25 observations) (ASTM 1995, 54).

Central line $= \bar{x}$

$$\text{Upper Control Limit} = \bar{x} + 3 \frac{\bar{s}}{\sqrt{n - 0.05}}$$

$$\text{Lower Control Limit} = \bar{x} - 3 \frac{\bar{s}}{\sqrt{n - 0.05}}$$

where

$\bar{x}$ = average number of nonconformities
n = number of cases in the sample
$\bar{s}$ = average standard deviation

Equation 7: P Charts

Fraction of Nonconforming Parts (ASTM 1995, 58).

Central line $= \bar{p}$

$$\text{Upper Control Limit} = \bar{p} + 3 \sqrt{\frac{\bar{p}(1-\bar{p})}{N}}$$

$$\text{Lower Control Limit} = \bar{p} - 3 \sqrt{\frac{\bar{p}(1-\bar{p})}{N}}$$

where

$\bar{p}$ = average proportion of cases
N = number of cases

Equation 8: Np Charts

Number of Nonconforming Units (ASTM 1995, 58).

Central line $= n\bar{p}$

Upper Control Limit $= n\bar{p} + 3\sqrt{n\bar{p}}$

Lower Control Limit $= n\bar{p} - 3\sqrt{n\bar{p}}$

where

p = average fraction of cases
n = number of cases in the sample

Equation 9: U Charts

Nonconforming Per Unit (ASTM 1995, 59).

Central line $= \bar{u}$

Upper Control Limit $= \bar{u} + 3\sqrt{\dfrac{\bar{u}}{N}}$

Lower Control Limit $= \bar{u} - 3\sqrt{\dfrac{\bar{u}}{N}}$

where

$\bar{u}$ = number of nonconformities per unit
N = number of cases in the sample

Equation 10: C Charts

Number of Nonconformities (ASTM 1995, 61).

Central line $= \bar{c}$

Upper Control Limit $= \bar{c} + 3\sqrt{\bar{c}}$

Lower Control Limit $= \bar{c} - 3\sqrt{\bar{c}}$

where

$\bar{c}$ = average number of nonconformities

There are a variety of other charts and graphs that can be used to monitor safety performance in addition to the control charts used in statistical process control. Some of the more common types include cumulative frequency distributions and Pareto charts.

CUMULATIVE FREQUENCY DISTRIBUTIONS

Cumulative frequency distributions are used to display the frequency of observations having a value greater than or less than a particular scale value (ASTM 1995, 10). Cumulative frequency distributions can be displayed in a table format or graphically. Table 1 represents a cumulative frequency distribution in tabular format. Using this table, it is possible to determine the number of cases above or below a particular category. To interpret this table, 10 subjects reported the presence of one CTD symptom, 15 subjects reported two or fewer CTD symptoms, 17 subjects reported 3 or fewer CTD symptoms, and so on. The entire sample consisted of 22 subjects reporting 5 or fewer CTD symptoms. By using the cumulative frequency distribution table, you can readily see that approximately 77.3 percent of the subjects reported 3 or fewer CTD symptoms.

Thus far, the control charts presented assume a normal distribution, meaning that the data can be expected to follow a bell-shaped curve. This may not always be the case. Two other common distributions that data can also follow are the Poisson distribution and the binomial distribution. If the data collected for the performance measure are assumed to follow either of these distributions, then the appropriate control should be used.

TABLE 1

	Cumulative Frequency Distribution Table	
Number of Reported CTD Symptoms	Number of Subjects Reporting CTD Symptoms Less than the Given Values	Percentage of Subjects Reporting CTD Symptoms Less than the Given Values
1	10	45.5%
2	15	68.2%
3	17	77.3%
4	19	86.4%
5	22	100.0%

POISSON DISTRIBUTIONS

Poisson distributions can be expected when the sample size is very large and the probability of the occurrence of an event of interest is very small. A good approximation to use in determining if the Poisson distribution is appropriate is when the sample size is greater than 50 and the probability of the occurrence of an event is less than 0.20 (Hays 1988, 145). In safety, many accident measures assume a Poisson distribution. For example, the probability of an accident

CASE STUDY

Example Poisson Distribution

A safety manager collected data on the number of damaged fire extinguishers throughout the entire company. He found that the overall probability of finding a damaged extinguisher is 0.003. However, at one facility, which has 500 extinguishers, he found 4 damaged extinguishers. What is the probability of finding this many damaged extinguishers in one facility?

$$p(x,m) = \frac{2.718^{-1.5} 1.5^4}{4!} = 0.05$$

where

$e = 2.718$
$m = 500 \times 0.003 = 1.5$
$x = 4$.

Therefore, based on prior experience with damaged fire extinguishers, the probability of finding 4 damaged extinguishers out of the 500 is 0.05, or a 5 percent chance.

---

can assume to follow a Poisson distribution when the frequency of accidents is small and the total exposure is large. The formula for calculating a Poisson probability is:

$$p(x,m) = \frac{e^{-m} m^x}{x!}$$

where

$e = 2.718$
m = expected number of events (number in sample × probability of one event)
x = number of events.

BINOMIAL DISTRIBUTIONS

A variable follows a *binomial distribution* when there are two possible outcomes. As with the Poisson and normal distributions, when collecting data for a performance measure that has two possible outcomes, the appropriate control chart based upon the binomial distribution must be used. The formula for calculating a binomial probability is:

$$p(x=r,n,p) = \frac{n!}{r!(n-r)!} p^r (1-p)^{n-r}$$

where

n = number of cases in sample
r = number of cases of interest
p = probability of one event.

PARETO CHARTS

The *Pareto Principle* states that in any population that contributes to a common effect, a relatively few of the contributors—the vital few—account for the bulk of the effect (Juran and Godfrey 1999, 5.20). The principle applies widely in human affairs. Like their tabular counterparts, each of these Pareto charts contains three elements (Juran and Godfrey 1999, 5.20):

1. The contributors to the total effect, ranked by the magnitude of their contribution
2. The magnitude of the contribution of each, expressed numerically and as a percentage of the total
3. The cumulative percentage of total contribution of the ranked contributors.

CASE STUDY

Example Binomial Distribution

It was determined that the probability of a worker from a particular construction company being trained on fall protection was 94 percent (6 percent are untrained). During an audit at one of the construction sites, a safety technician found that, out of 400 laborers interviewed, 10 stated they had not been trained in fall protection. What is the probability of finding this many untrained laborers?

$$p(x=r,n,p) = \frac{400!}{10!(400-10)!} 0.06^{10}(1-0.06)^{400-10}$$
$$= 0.0005$$

where

$n = 400$
$r = 10$
$p = 0.06$.

When selected at random, 6 percent of the workers could be expected to be untrained. With a workforce of 400 laborers, one could expect to find 24 untrained laborers. However, only 10 were identified which is considerably less than expected. The probability of this occurring is 0.0005. Based on these results, the safety technician can conclude that there are significantly fewer untrained workers on this job site than can be expected due to chance.

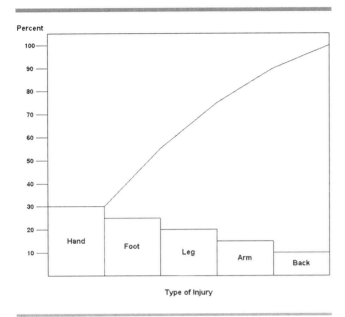

FIGURE 2. Sample Pareto chart

In the displayed Pareto chart (see Figure 2), the frequencies of injuries reported by body part were collected. The injury categories are arranged in order from highest to lowest percentage of cases. The bars represent the percentage of reported cases by category, while the line represents the cumulative percentage of reported cases. As can be inferred from Figure 2, hand injuries account for 30 percent of the reported cases, hand and foot injuries account for 55 percent, and so on. The Pareto analysis indicates that, in this example, hand injuries account for the greatest percentage of reported cases.

Using Existing Criteria to Develop Performance Measures

Rather than develop unique performance measures, it is common for safety professionals to use existing performance measures developed by other entities. Examples of sources of safety performance measures used in industry include OSHA's Voluntary Protection Program Criteria, OSHA's Safety and Health Program Assessment Worksheet (Revised OSHA Form 33) (which was discussed previously in this chapter), ISO 9000:2000 Standards, ISO 14000 Standards, and Six Sigma methodologies. Each of these criteria have been used in industry as a means to measure and improve safety performance in the workplace.

Voluntary Protection Program Criteria

On July 2, 1982, OSHA announced the establishment of the Voluntary Protection Programs (VPP) to recognize and promote effective work-site-based safety and health management systems (OSHA 2008a). VPP is OSHA's official recognition of the outstanding efforts of employers and employees who have created exemplary work-site safety and health management systems. The following principles are embodied in the Voluntary Protection Programs (OSHA 2008a):

- **Voluntarism.** Participation in VPP is strictly voluntary. The applicant who wishes to participate freely submits information to OSHA on its safety and health management system and opens itself to agency review.
- **Cooperation.** OSHA has long recognized that a balanced, multifaceted approach is the best way to accomplish the goals of the OSH Act. VPP's emphasis on trust and cooperation between OSHA, the employer, employees, and employees' representatives complements the agency's enforcement activity, but does not take its place. VPP staff and participating sites work together to resolve any safety and health problems that may arise. This partnership enables the agency to remove participating sites from programmed inspection lists, allowing OSHA to focus its inspection resources on establishments in greater need of agency oversight and intervention. However, OSHA continues to investigate valid employee safety and health complaints, fatalities, catastrophes, and other significant events at VPP participant sites.
- **A Systems Approach.** Compliance with the OSH Act and all applicable OSHA requirements is only the starting point for VPP sites. VPP participants develop and implement systems to effectively identify, evaluate, prevent, and control occupational hazards so that injuries and illnesses to employees are prevented. Star sites, in particular, are often on the leading edge of hazard prevention methods and technology. As a result, VPP work sites serve as models of

safety and health excellence, demonstrating the benefits of a systems approach to worker protection. NOTE: Federal agencies participating in VPP also must comply with Executive Order 12196 and 29 CFR 1960, in addition to Section 19 of the OSH Act.
- **Model Work Sites for Safety and Health.** OSHA selects VPP participants based on their written safety and health management system, the effective implementation of this system over time, and their performance in meeting VPP requirements. Not all work sites are appropriate candidates for VPP. At qualifying sites, all personnel are involved in the effort to maintain rigorous, detailed attention to safety and health. VPP participants often mentor other work sites interested in improving safety and health, participate in safety and health outreach and training initiatives, and provide OSHA with input on proposed policies and standards. They also share best practices and promote excellence in safety and health in their industries and communities.
- **Continuous Improvement.** VPP participants must demonstrate continuous improvement in the operation and impact of their safety and health management systems. Annual VPP self-evaluations help participants measure success, identify areas needing improvement, and determine needed changes. OSHA on-site evaluation teams verify this improvement.
- **Employee and Employer Rights.** Participation in VPP does not diminish employee and employer rights and responsibilities under the OSH Act and, for Federal agencies, under 29 CFR 1960 as well.

The categories of VPP participation consist of (OSHA 2008a):

- **Star Program.** The Star Program recognizes the safety and health excellence of work sites where workers are successfully protected from fatality, injury, and illness by the implementation of comprehensive and effective workplace safety and health management systems. These work sites are self-sufficient in identifying and controlling workplace hazards.
- **Merit Program.** The Merit Program recognizes work sites that have good safety and health management systems and that show the willingness, commitment, and ability to achieve site-specific goals that will qualify them for Star participation.
- **Star Demonstration Program.** The Star Demonstration Program recognizes work sites that have Star quality safety and health management systems that differ in some significant fashion from the VPP model and thus do not meet current Star requirements. A Star Demonstration Program tests this alternative approach to protecting workers to determine if it is as protective as current Star requirements.

OHSAS 18001:2007

OHSAS 18001:2007 is an international standard for safety management systems. The standard was developed by a selection of leading trade bodies and international standards and certification bodies to address a gap where no third-party certifiable international standard exists (BSI 2010).

While OHSAS 18001:2007 is not recognized by the International Organization for Standardization (ISO), it is compatible with ISO 9001 and ISO 14001 (BSI 2010).

The following key areas are addressed by this international standard:

- planning for hazard identification, risk assessment, and risk control
- OHSAS management program
- structure and responsibility
- training, awareness, and competence
- consultation and communication

- operational control
- emergency preparedness and response
- performance measuring, monitoring, and improvement

OHSAS 18002 provides the guidelines for the implementation of OHSAS 18001.

ISO STANDARDS

The International Organization for Standardization, commonly referred to as ISO, is a nongovernmental organization founded in 1947 with its headquarters in Geneva, Switzerland (ISO 2004a). Its mission is to promote the development of standardization and related activities in the world. While the organization develops hundreds of standards, two families of standards—ISO 9001:2008 and ISO 14000 in particular—have had a dramatic impact on how safety programs are implemented and how performance is measured in the workplace.

ISO 9001:2008

The ISO 9001:2008 standard represents an international consensus on good quality management practices (ISO 2010a). Revisions to the ISO 9000 family of standards in 2008 resulted in a consolidation of ISO 9001, 9002, and 9003 into the ISO 9001:2008.

The ISO 9001:2008 standard provides guidance on the development of a quality management system, management responsibilities, resource management, product realization, and measurement and improvement processes.

Safety professionals have found it advantageous for organizations that have adopted the ISO standards as their quality improvement process to incorporate safety into the ISO management processes.

ISO 9004:2009

In addition to ISO 9001:2008, ISO 9004:2009 provides guidance to organizations, supporting the achievement of sustained success through a quality management approach. It is applicable to any organization, regardless of size, type, and activity (ISO 2010b).

ISO 14000

The ISO 14000 family of standards establishes the framework for environmental management systems. An environmental management system is designed to ensure an environmentally friendly product throughout the entire life cycle. Table 2 summarizes the various ISO 14000 standards that are included in the Environmental Management System and in the ISO 14000 Model.

Two standards in the ISO 14000 group that are of particular importance to the safety professional are ISO 14001:2004 and ISO 14004:2004. Both have recently undergone revisions. These standards establish the framework for an environmental management system (EMS). ISO 14001:2004 specifies the requirements for an EMS that provides a framework for an organization to control the environmental impact of its activities, products, and services, and to continually improve its environmental performance.

ISO 14001:2004 specifies requirements for an EMS that enables an organization to develop and implement policies and objectives that take into account the legal and other requirements the organization subscribes to, and information about significant environmental aspects. It applies to the environmental aspects that the organization has identified as being those it can control and those that it can influence. It does not itself state specific environmental performance criteria. ISO 14001:2004 is applicable to any organization that wishes to establish, implement, maintain, and improve an environmental management system, to assure itself of conformity with its stated environmental policy, and to demonstrate conformity with ISO 14001:2004 by (ISO 2004b):

- making a self-determination and a self-declaration
- seeking confirmation of its conformance by parties having an interest in the organization, such as customers

TABLE 2

The ISO 14000 Family of International Standards (ISO 2009)

Designation	Publication Title
ISO 14001:2004	Environmental management systems – Requirements with guidance for use
ISO 14004:2004	Environmental management systems – General guidelines on principles, systems, and supporting techniques
ISO/DIS 14005	Environmental management systems – Guidelines for the phased in implementation of an environmental management system
ISO/CD 14006	Environmental management systems – Guidelines on ecosystems
ISO 14015:2001	Environmental management – Environmental assessment of sites and organizations (EASO)
ISO 14020:2000	Environmental labels and declarations – General principles
ISO 14021:1999	Environmental labels and declarations – Self-declared environmental claims (Type II environmental labeling)
ISO 14024:1999	Environmental labels and declarations – Type I environmental labeling – Principles and procedures
ISO 14025:2006	Environmental labels and declarations – Type III environmental declarations – Principles and procedures
ISO 14031:1999	Environmental management – Environmental performance evaluation guidelines
ISO/AWI 14033	Environmental management – Quantitative environmental information – Guidelines and examples
ISO 14040:2006	Environmental management – Life cycle assessment – Principles and framework
ISO 14044: 2006	Environmental management – Life cycle assessment – Requirements and guidelines
ISO/WD 14045	Eco-efficiency assessment – Principles and requirements
ISO/TR 14047: 2003	Environmental management – Life cycle assessment – Examples of application of ISO 14042
ISO/TS 14048:2002	Environmental management – Life cycle assessment – Data documentation format
ISO/TR 14049:2000	Environmental management – Life cycle assessment – Examples of application of ISO 14041 to goal and scope definition and inventory analysis
ISO 14050:2009	Environmental management – Vocabulary
ISO/CD 14051	Environmental management – Material flow cost accounting – General principles and framework
ISO/TR 14062:2002	Environmental management – Integrating environmental aspects into product design and development
ISO 14063: 2006	Environmental management – Environmental communication – Guidelines and examples
ISO 14064-1:2006	Greenhouse gases – Part 1: Specification with guidance at the organization level for quantification and reporting of greenhouse gas emissions and removals
ISO 14064-2:2006	Greenhouse gases – Part 2: Specification with guidance at the project level for the quantification, monitoring and reporting of greenhouse gas emission reductions or and removal enhancements
ISO 14064-3:2006	Greenhouse gases – Part 3: Specification for the validation and verification of greenhouse gas assertions
ISO 14065: 2007	Greenhouse gases – Requirements for greenhouse gas validation and verification bodies for use in accreditation or other forms of recognition
ISO/CD 14066	Greenhouse gases – Competency requirements for greenhouse gas validators and verifiers document
ISO/WD 14067-1	Carbon footprint of products – Part 1: Quantification
ISO/WD 14067-2	Carbon footprint of products – Part 2: Communication
ISO/AWI 14069	GHG-Quantification and reporting of GHG emissions and organizations – Guidance for the application of ISO 14064-1
ISO 19011:2002	Guidelines for quality and/or environmental management systems auditing
ISO Guide 64:2008	Guide for the inclusion of environmental aspects in product standards

Key to Abbreviations: AWI = Approved Work Item., CD = Committee Draft, DIS = Draft International Standard, TR = Technical Report, WD = Working Draft

- seeking confirmation of its self-declaration by a party external to the organization
- seeking certification/registration of its environmental management system by an external organization.

ISO 14004:2004 provides guidance on the establishment, implementation, maintenance, and improvement of an environmental management system and its coordination with other management systems. The guidelines in ISO 14004:2004 are applicable to any organization, regardless of its size, type, location, or level of maturity. While the guidelines in ISO 14004:2004 are consistent with the ISO 14001:2004 environmental management system model, they are not intended to provide interpretations of the requirements of ISO 14001:2004 (ISO 2004b).

REFERENCES

Al-Mutairi, A., and J. M. Haight. 2009. "Predicting Incident Rates." *Professional Safety* 54 (9):40–48.

American Industrial Hygiene Association (AIHA). 2005. *American National Standard for Occupational Safety and Health Management Systems*. Fairfax, VA: AIHA.

American Society for Testing and Materials (ASTM). 1995. *Manual on Presentation of Data and Control Charts*. Philadelphia, PA: ASTM.

British Standards Institution (BSI). 2010. *BS OHSAS 18001: Occupational Health and Safety* (accessed November 21, 2010). www.bsigroup.com/en/Assessment-and-certification-services/management-systems/Standards-and-Schemes/BSOHSAS-18001/

Hays, William. 1988. *Statistics*. Orlando, FL: Holt, Rinehart and Winston.

International Organization for Standardization (ISO). 2004a. *Overview of the ISO System*. www.iso.org/iso/en/aboutiso/introduction/index.html#two

———. 2004b. *ISO 14001: 2004: Environmental Management Systems—Requirements for Guidance and Use*. www.iso.org/iso/en/CatalogueDetailPage.CatalogueDetail?CSNUMBER=31807&ICS1=13&ICS2=20&ICS3=10&scopelist=ALL

———. 2009. *Environmental Management: The ISO 14000 Family of International Standards*. Geneva, Switzerland: ISO.

———. 2010a. *ISO 9000 Essentials*. www.iso.org/iso/iso_catalogue/management_standards/iso_9000_iso_14000/iso_9000_essentials.htm

———. 2010b. New edition of *ISO 9004* maps out the path forward to "sustained success." www.iso.org/iso/pressrelease.htm?refid=Ref1263

Iyer, P. S., J. M. Haight, E. Del Castillo, B. W. Tink, and P. W. Hawkins. 2004. "Intervention Effectiveness Research: Understanding and Optimizing Industrial Safety Programs Using Leading Indicators." *Chemical Health & Safety* 11(2):9–20.

Juran, Joseph M., and A Blanton Godfrey. 1999. *Juran's Quality Handbook*. New York: McGraw-Hill.

Lack, Richard W. 2002. *Safety, Health, and Asset Protection*. Boca Raton, FL: Lewis Publishers.

Mohamed, Sherif. 2002. "Safety Climate in Construction Site Environments." *Journal of Construction Engineering & Management* 128(5):375–385.

Occupational Safety and Health Administration (OSHA). 1998. *Publication 2552: Training Requirements in OSHA Standards and Training Guidelines*. Washington, D.C.: OSHA.

———. 2004. *Safety and Health Management eTool*. www.osha.gov/SLTC/etools/safetyhealth/components.html

———. 2008a. *Voluntary Protection Programs (VPP): Policies and Procedures Manual* (OSHA Directive Number CSP 03-01-003). Washington, D.C.: OSHA.

———. 2008b. *Consultation Policies and Procedures Manual* (OSHA Directive Number CSP 02-00-002). Washington, D.C.: OSHA.

Petersen, Dan. 1996. *Analyzing Safety System Effectiveness*. New York: Van Nostrand Reinhold.

Pierce, F. David. 1995. *Total Quality for Safety and Health Professionals*. Rockville, MD: Government Institutes.

Shewhart, Walter. 1931. *The Economic Control of Quality of Manufactured Product*. New York: Van Nostrand and Company.

Swartz, George. 2000. *Safety Culture and Effective Safety Management*. Chicago, IL: National Safety Council.

Weinstein, Michael B. 1997. *Total Quality Safety Management and Auditing*. Boca Raton, FL: CRC Press LLC.

7

BEST PRACTICES

Linda Rowley

LEARNING OBJECTIVES

- Describe the concepts and framework related to the management of safety engineering work.

- Describe the role of safety in the system life cycle.

- Identify some of the professional organizations, federal agencies, resources, references, and publications for safety engineering practices and management.

- Define various management approaches used for major systems.

THIS CHAPTER PROVIDES an overview of the best practices in the management of safety engineering work and includes a general overview of basic safety engineering concepts and principles, references to regulations and standards, and management practices. In this chapter, the management elements in a comprehensive framework are grouped and summarized to provide a general description of safety engineering and its interfaces with the organization. Best practices in safety engineering work are developed from research, lessons learned in past experience, and examining successful models. In one perspective, a best practice could be an applied methodology or a standard practice in the development of a system. From another perspective, best practices are constantly evolving, where both process and product benefit from keeping two goals at the forefront: safety and efficiency.

The field of engineering is involved with the application of mathematical, physical, and scientific principles used to plan, design, and build systems. There are traditional fields of engineering, such as civil, electrical, industrial, mechanical, and chemical, as well as specialized fields such as information systems technology, aerospace, nuclear, medical, construction, mining, and safety. Safety engineering is a specialized field that applies engineering principles, criteria, and techniques to identify and eliminate hazards and manage risk.

System safety is a management framework that facilitate safety engineering work, including the review and decision-making process of identifying, evaluating, and controlling hazards, and managing risks from concept through system disposal. Although the identification of hazards, risks, and corrective actions can be applied to existing systems, system safety engineering is ideally applied to new designs, facilities, and processes. Including safety as part of the system design is considerably more efficient and cost effective than retrofitting or adding safety features to a system.

The terms safety engineering and system safety are often used interchangeably, but there is a subtle difference. *Safety engineering* draws upon analytical techniques and tools, such as fault tree analysis and failure modes and effects analysis, to enhance the design and operation of a system. *System safety* is the holistic framework that facilitates the management of hazards for the life cycle of a system or product. Managing safety helps to identify and monitor hazards; increase efficiency; and control costs to operate and maintain a system safely in its intended environment with minimal risk to users, the general public, property, and the environment.

According to Clemens and Simmons (2002):

> System safety has two characteristics: (1) it is a doctrine of management practice that mandates that hazards be found and risks controlled; and (2) it is a collection of analytical approaches with which to practice the doctrine. Systems are analyzed to identify their hazards, and those hazards are assessed for a single reason: to support management decision making.

This chapter focuses on the framework, process, and management of safety engineering work, identifying the principles and best practices in management. As Leveson states (2004):

> Whereas industrial (occupational) safety models focus on unsafe acts or conditions, classic system-safety models instead look at what went wrong with the system's operation or organization to allow the accident to take place.

Best practices are dynamic organizational learning systems achieved by benefiting from previous experiences, acquiring new information through research or reviewing the literature, benchmarking and measuring performance, and committing to learning and continuous improvement. Various organizational examples in this chapter highlight the structure and principles of common and accepted practices in safety engineering management, references to governmental requirements and professional organizations, and publications as resources for lessons learned and best practices.

The management system structures the process for managing safety engineering activities. The framework, including leadership, the formal structure, communications, training, and risk-based decision making, is critical in managing safety engineering work.

SAFETY ENGINEERING IN THE SYSTEM LIFE CYCLE

System safety is a comprehensive approach for integrating safety as part of the design—and implementing requirements throughout other phases—in the life cycle of a system, product, process, or facility. The primary function of system safety is to identify and control hazards in each phase of the life cycle, from concept through decommissioning and disposal. In system development, anticipating potential hazards and conditions is a key aspect of safety engineering work. It is a challenge to anticipate the hazards of a system before it is developed; however, safety engineering activities designed into each phase promote a systematic process of anticipating and identifying hazards as the system is developed. According to Brauer (1990), "System safety is . . . the systematic, forward-looking identification and control of hazards. . . ." Safety engineers assess the existing and potential conditions that could affect a system.

There are four major phases of development in a system life cycle: (1) concept, (2) system development, (3) production and deployment, and (4) sustainment and disposal.

Each phase includes safety engineering tasks that result in a formal decision about proceeding to the next phase. A system-safety management plan should be developed during the concept phase in order to design safety into the system and maintain it throughout the system's life. Incorporating system safety early in development increases the probability that hazards can be addressed more economically and with greater efficiency. A formal decision earmarks the acceptance of risk(s) to that point of development or operation.

Risk can have many different meanings (Main 2004), and the perception of risk is often related to the perspective of the assessor. For example, risk perceived from a fire or health perspective would be different than risk perceived from a financial or security perspective. In safety, risk is based on the probability and severity of a hazard or unsafe condition. Risk is commonly categorized by high, medium, and low risk where quantitative and qualitative data support the categorization of risk. There are variations on a risk model that might include additional risk categories such as negligible, very low, and extremely high. The risk-management matrix provides a consistent method of evaluating and managing hazards for that system, or could be used as an organizationwide risk-assessment method. Three fundamental best practices emerge in managing safety engineering work: (1) a common definition of risk and methodology for assessing hazards; (2) collaboration, allowing the integration of safety engineering into each phase of design and development; and (3) risk decisions for each hazard.

Incorporating safety early in system development increases the probability that hazards will be addressed in each development phase. The opportunities for timely and cost-effective solutions are addressed in early design stages rather than in later phases that potentially have significant impact on development, schedules, and costs. According to Beohm (Marshall 2000), "Safety engineers and professionals should be involved in the conceptual stages of any project design until its final commissioning to ensure that safety is an integral part of that product."

The management or project plan identifies safety engineering requirements for the project, usually for the following activities:

- design, construction, operation, and modification of facilities and structures
- new product design, development, or acquisition
- new process or change
- redesign of a facility or process

Many organizations have established policies and requirements for system safety in the concept stage, from which a requirements document can be formulated. System-safety requirements also apply to subsystems and components, which are assessed both individually and wholly to determine interoperability and interfaces with other components within the entire system. Each phase of development ideally includes risk decision making at the system and subsystem levels.

The management structure and processes vary among organizations, but the commonalities are consideration of safety in the planning stage and a deliberate process of evaluation and decision making. A best practice in managing safety engineering work is a structured management model that facilitates the process for early and continuing involvement of safety engineering.

Management tools, such as the National Aeronautics and Space Administration's (NASA) *Systems Engineering Handbook* (NASA 2007) and *System Safety Handbook* (NASA 1999) provide a systematic approach to managing systems governed by NASA from concept through final disposition. The system-safety framework is the premise NASA established to manage safety engineering work in its systems and facilities. According to NASA, "The System Safety Program Plan (SSPP) is the most important element in implementing a system safety program" (NASA 1999, p. 4-2). Safety engineering work is included in each phase of the life cycle of a system or facility. The management framework for each project identifies key activities and responsibilities for each phase of the life cycle. The major components of the SSPP are:

- Establish the objectives.
- Define responsibilities for performing safety tasks.
- Identify interfaces with other disciplines and organizations.
- Define the tasks to achieve the objectives.
- Specify the management review process and controls.
- Determine the methods for conducting safety analyses.

Federal government system-safety requirements standardize the planned approach for safety engineering work within its organizations, schedules, activities, and decision making. For example, the management process for the Department of Defense (DOD), found in Military Standard 882E, is the structure for managing safety engineering work from concept through all phases to disposal (DOD 2000). There are eight requirements used for each system:

1. Establish the system-safety approach. This is the management plan that defines the

requirements and milestones for how safety will be integrated in system developments and how risks will be identified, communicated, accepted, and tracked.

2. Identify hazards through safety engineering methods that could occur in hardware, software, operations and use, and the environment.
3. Assess the risk in terms of severity and probability to determine the potential impact of each hazard. In MIL-STD 882E, a standard risk-assessment matrix is used to provide consistency in risk analyses and decisions.
4. Identify risk mitigation measures and alternatives. The order of priority for identifying alternatives and reducing risk is:
    (a) design safety into the system
    (b) use safety devices
    (c) provide warning devices
    (d) use procedures and training to reduce risk
5. Reduce unacceptable risks, using the order of prioritization.
6. Verify risk control measures through testing, inspection, and other methods, that show the control measures work as intended and do not introduce other unanticipated hazards.
7. Establish formal risk acceptance at an appropriate level of management.
8. Track all hazards throughout the life cycle of a system, including the hazards of disposal.

This management approach can be universally used for any industrial application of safety engineering work.

The National Institute of Building Sciences (NIBS) promotes an integrated team approach to buildings and their components.

These examples demonstrate a structure of established regulatory and industry procedures and system safety models that facilitate the management of safety engineering work. The formal structure provides a path that incorporates collaboration and integration of safety engineering activities, especially important in complex systems that have high risks. Management tools, such as plans and schedules, should be tailored for specific systems. One management tool that is often used is a project plan that schedules various activities and timelines in each phase of development. Project management principles are universal tools that can be applied to any system, building, process, or product.

Safety engineering work is facilitated through sound management principles that provide the framework of the management system, which includes the following activities:

- *Planning:* Beginning during the concept phase, it is included for each phase of system development, and is modified for continuous improvement throughout the life cycle of the system.
- *Organizing:* Management structure, system-safety process, people, resources, and activities.
- *Budgeting:* The costs of system development, testing, and operations of the system from concept through final disposition, including maintenance, replacement, and decommissioning.
- *Communicating:* A formal structure of written documents concerning the policies, procedures, development, hazards, and risk decisions that supplement informal communications, such as problem solving.
- *Decision making:* A formal process aligned within the system-safety program that determines the criteria for decision making, including the decision authority levels, actions taken, timelines, and risk decision(s) (acceptance or alternatives).
- *Change management:* The written process and implementation of changes within the organization or in the system life cycle.
- *Implementing:* A formal management structure that defines roles and responsibilities.
- *Evaluating:* Both informal and formal methods of monitoring and assessing safety engineering work, activities, decisions, and the efficiency of the system-safety process.

While the management plan does not usually determine the analytical methods for the implementation of safety engineering at a project level, it offers a structure for integrating safety engineering into the process in the context of evaluating risks and making decisions at strategic points in a system's development.

Formal decision making is a key component of system safety. A risk-assessment matrix, which includes the identification, evaluation, and prioritization of hazards, is used to make risk decisions.

Risk management is concerned with three primary goals: (1) managing hazards, (2) designing an efficient system, and (3) cost-efficiency.

ORGANIZATIONAL MANAGEMENT AND SAFETY ENGINEERING

Leadership and Organizational Management

Leadership is a fundamental component of the organizational and safety culture. A written policy statement, usually in the form of the organizational vision, mission, policies, and procedures, sets the direction for the strategic plans, guiding principles, and expectations for the organization. A written statement of the executive leadership's commitment, by itself, does not determine the culture, but it is a major influence in shaping the management systems and organizational culture. The leaders of the organization—top executive(s), board of directors, or owner(s)—must communicate, by word and action, a commitment to safety. For example, the National Safety Council (NSC) uses the phrase "the 3 E's of Safety" (engineering, education, and enforcement). The first "E," engineering, is the preferred choice in eliminating or mitigating hazards in a system. Where it is not feasible to engineer hazards from the design or operation of a system, education is used as an alternative and supplementing method to inform users of hazards. Lastly, the third "E," enforcement of the requirements, includes supervision and monitoring.

At the organizational level and in the system-safety plan, it is important to articulate organizational values and priorities. "The causes of accidents are frequently, if not always, rooted in the organization—its culture, management, and structure" (Leveson et al. 2004, 1). In studies of companies with exceptional safety performance, four common characteristics were identified (Marshall 2000):

1. Strong management commitment
2. Efficient hazard identification
3. Efficient employee communication and involvement
4. Integration of safety into the larger management system

For example, NASA leadership has established the following high-level safety objectives for managing hazards (NASA 2008):

1. Protect public health
2. Protect workforce health
3. Protect the environment
4. Protect a program (systems and infrastructures needed to execute a mission) and public assets

NASA asserts that (NASA 2008, 5):

> . . . to properly support key design and operational decisions, it is necessary that design and operational alternatives are analyzed not only with respect to their impact on the mission's technical and programmatic objectives, but also with respect to their impact on these high-level safety objectives.

Hazard identification is the primary activity in safety engineering work. While the organizational culture influences the priorities and decisions made during system development, management has a direct relationship with safety activities, resources, schedules, and decisions. The management framework incorporates and addresses resource allocation and other management issues that arise from four sources of failure (Haimes 2009, 25):

1. Hardware failure
2. Software failure
3. Organizational failure
4. Human failure

The system-safety program provides a mechanism for managing safety engineering work by aligning activities and resources within a planned timeframe.

There are likely to be multiple, and sometimes competing, perspectives and objectives. An essential element of system safety involves making decisions that eliminate hazards or reduce the frequency and/or consequences of accidents involving hazards to an acceptable level by introducing hazard control measures and modifying system design and/or procedures. The organization of the system-safety program is critical to:

1. Integrating safety engineering functions with the development of a system

2. Making risk decisions at each phase of development

According to Haimes (2009, 3), "Uncertainty colors the decision-making process regardless of . . ." the involvement of one or more parties; economic, sociopolitical, or geographical issues; science and technology; power and stakeholder involvement. Decision tools are helpful in establishing the criteria to make decisions about whether hazards are eliminated, mitigated, transferred, or accepted. Consequently, risk decision authority is an important element of the system-safety process.

In a report about NASA's safety culture, Leveson wrote that culture is embedded in routine, everyday activities, as well as in the organizational structure (2004): ". . . all aspects of the culture that affect safety must be engineered to be in alignment with the organizational safety principles." Common engineering methods, such as fault tree analysis (FTA) and failure modes and effects analysis (FMEA), are components of a system safety program within a comprehensive organizational safety structure. Leveson (2004) further states:

> Management, resources, capabilities, and culture are intertwined, and trying to change the culture without changing the environment within which the culture operates is doomed to failure. At the same time, simply changing the organizational structures—including policies, goals, missions, job descriptions, and standardized operating procedures related to safety—may lower risk over the short term, but superficial fixes that do not address the set of shared values and social norms are very likely to be undone over time.

A safety culture is complex; however, the organization's management structure and practices are significant factors in influencing the safety culture.

Culture is a pattern of shared basic assumptions that members of a group perceive, think, and feel in solving problems (Schein 1993). According to Schein, who is credited with coining the term *corporate culture*, a culture is comprised of three main components: (1) assumptions, (2) espoused values, and (3) artifacts and behaviors. Safety engineering work is largely influenced by the culture of the organization. The management structure, including goals, policies, plans, procedures, resources, scheduling, and other tools, are artifacts—visible and tangible activities that influence the culture. The three components are interdependent entities that collectively create an environment that affects individuals, work, and the organization.

Safety culture is a term that was created in 1989 in a report by the International Atomic Energy Agency (IAEA) after the Chernobyl nuclear accident (IAEA 1989). According to the IAEA, a safety culture is the characteristics and attitudes of the organization and individuals concerning the perception of the value and prioritization of safety. Similarly, another definition was created after the NASA Columbia accident as "the subset of organizational culture that reflects the general attitude toward and approaches to safety and risk management" (Leveson et al. 2005, 5). A safety culture is complex and unique within each organization. Creating and sustaining a safety culture is challenging, especially so in organizations with diversity in geographical locations, contractor support activities, multiple operations, and other variables. The overall organizational safety culture is less likely to be changed, or developed, as easily as the management structure and resources.

However, the organization's management structure and practices are significant factors in influencing the safety culture. The management framework has a direct effect on the process of performing safety engineering work. In addition, the management framework has a dual effect, indirectly affecting the culture, as well as being indirectly affected by the culture.

The best practice for optimizing safety engineering work is to create an organizational culture that embraces safety. This practice takes a multipronged approach in planning and implementing the three components that create an environment where safety engineering work is esteemed as a critical function in developing a product, facility, or system. Safety engineers and managers may have divergent perspectives in optimal solutions for system safety. As stated by Main (2004, 381):

> Engineers typically have a strong passion for technical accuracy . . . conversely, managers and implementers tend to focus on value and effectiveness and often lack passion for technical accuracy. Managers may willingly accept some inaccurate but effective solutions that are easily implemented.

The best approach for a particular company is the method that works best in the organizational culture and design processes (Main 2004, 423).

PLANS, RESOURCES, AND COSTS

The primary purposes of safety engineering work are to prevent loss and optimize efficiency. Management of safety engineering work involves planning, budgeting, scheduling, and weighing risks in each phase of the system. Establishing criteria for decisions during the planning phases of system safety creates an objective method for weighing the costs and risks in the management of the hazards. Planning includes establishing guidelines and parameters within given cost and time constraints. Turner (2003) cautions that "the most important question to ask of any practice is the size of the bill."

The budget for any system, facility, or product includes cost and time consideration on two levels. First, there must be adequate resources allocated for safety engineering during each phase of development. For example, in the concept phase, a preliminary hazard analysis serves as a basis to create a system-safety plan. In addition to establishing milestones in each phase of system development, there is an opportunity to minimize costs and maximize efficiency by considering other disciplines that would enhance working groups.

Second, cost is part of the risk evaluation and decision for each hazard. It is likely that most people would be able to use more resources, but it is usually not efficient or feasible to eliminate all hazards in a system. Using the principles of risk management, risks may be transferred, controlled, or accepted. Cost and schedules are considered in a risk decision, but should not be the primary determinants in system development. Errors, time pressures, inadequately trained personnel, and unscheduled downtimes can have adverse effects on costs and schedules, indirectly contributing to the potential for accidents or oversights and elevating system costs.

There are numerous examples where resources, in terms of budget and time, have been cited as contributing factors in a system catastrophe. Leveson and Cutcher-Gershenfeld (2004) noted that one of the lessons learned in the accident analysis of the Columbia accident was that "budget cuts without concomitant cuts in goals led to trying to do too much with too little" (Leveson and Cutcher-Gershenfeld 2004, 1). Another example of a catastrophic accident occurred at the BP Texas City refinery in Texas City, Texas, in 2005. Among numerous root and contributing causes of the accident, the U.S. Chemical Safety Board (CSB) report found that risks were oversimplified and BP failed to address serious safety hazards (CSB 2005, 186), concluding that "budget cuts impaired process safety performance" (CSB 2005, 210).

There is much to be learned from catastrophic accidents concerning the importance of plans and resources for safety engineering work. A best practice in managing safety engineering work is to develop a system-safety management plan that includes adequate resources for resolution of serious hazards.

FEDERAL REGULATIONS AND RESOURCES

System-safety requirements are mandated for critical systems that can adversely affect life, health, property, operations, and the environment. In addition to regulatory requirements, federal agencies that develop systems provide a framework and guidance for managing safety engineering work for the respective systems they develop. The framework and guidance would include such elements as policies, procedures, handbooks, responsibilities, tasks, and other such elements.

The following government organizations regulate and oversee the development, operation, and disposal of certain critical systems. Commercial firms and other contractors implement government regulations but do not regulate the affected industry. Government organizations and industries develop provisions to implement federal regulations and may exceed minimal standards.

The Code of Federal Regulations (CFR) represents the standards and requirements of executive departments and federal agencies that are generated from federal laws. For example, the Occupational Safety and Health Administration (OSHA), created by the Occupational Safety and Health Act of 1970, establishes standards for workplace safety and health, and is an administrative branch of the Department of Labor. The law establishes the requirements, while other organizations, such as the American National Standards Institute (ANSI), the Institute of Electrical and Electronics Engineers, Inc. (IEEE), and the American Society of Safety Engineers (ASSE), develop standards, practices, guidelines, and other tools organizations can use to comply with the requirements and industry practice.

Below are some of the federal government agencies and regulations for system-safety requirements:

- Department of Defense Instruction (DoDI): DoDI 5000.1, The Defense Acquisition System; DoDI 5000.2, Operation of the Defense Acquisition System; and Military Standard 882D, DoD Standard Practice for System Safety
- Department of Transportation, Federal Aviation Administration (FAA): 14 CFR 1, Aeronautics and Space, FAA *System Safety Handbook*
- Department of Transportation, National Highway Traffic Safety Administration (NHTSA): Federal Motor Vehicle Safety Standards (FMVSS) 49 CFR 5 and 301
- Environmental Protection Agency (EPA): 40 CFR 68, Chemical Accident Prevention Provisions
- Food and Drug Administration (FDA): 21 CFR 807.90, Pre-Market Notification for Medical Devices; Part 210, Current Good Manufacturing Practice in Manufacturing, Process, Packing, or Holding of Drugs; 21 CFR Part 211, Current Good Manufacturing Process for Finished Pharmaceuticals
- Nuclear Regulatory Commission (NRC): 10 CFR 70.62, Safety Program and Integrated Safety Analysis; 10 CFR 70.72, Facility Changes and Change Process
- Occupational Safety and Health Administration (OSHA): 29 CFR 1910.119, Process Safety Management; 29 CFR 1910.147, Control of Hazardous Energy
- United States Patent and Trademark Office: *Life Cycle Management Manual*

Many organizations use voluntary guidelines and industry consensus standards to enhance capability, maximize resources, and mitigate hazards, thereby minimizing the necessity of redesign for correcting hazards and reducing other costs that may adversely affect people, property, or the environment.

INDUSTRY STANDARDS AND PROFESSIONAL ORGANIZATIONS

In addition to the mandatory federal regulations, organizations advance industry standards, research, and current information through publications, guidelines, and practices developed by technical experts from government, industry, academia, and other stakeholders. Professional nonprofit organizations also publish timely peer-reviewed articles from their members. A few that publish professional journals in addition to other member services are listed below. These journals are excellent resources for standards and practices that have been developed and recommended by technical and scientific committees.

- The American National Standards Institute (ANSI), a nonprofit organization, coordinates and publishes voluntary standards that exceed regulatory requirements. Experts from industry, academia, and government work on technical committees to develop consensus standards. One of the recent publications, *ANSI Z-10, Occupational Health and Safety Management Systems*, was created through a partnership with the American Industrial Hygiene Association (AIHA). ANSI has numerous committees, subcommittees, and publications.
- The American Petroleum Institute (API) is the national trade association for the oil and natural gas industry and produces quality standards for the industry. One of its standards is *Specification for Quality Programs for the Petroleum, Petrochemical and Natural Gas Industry* (API 2007), which includes the requirements of the International Organization for Standardization (ISO) 9001 and additional quality assurance requirements for the oil and gas industry. The API does not regulate the industry; however, their guidance documents provide a systematic approach to integrating safety, health, and

environment considerations for equipment, products, and services in the global industry.
- The American Society of Safety Engineers (ASSE) is a nonprofit professional safety organization that advances research, technical publications, and other initiatives concerning protection of people, property, and the environment.
- The American Society for Testing and Materials (ASTM) is an international organization that publishes technical standards and specifications for materials, products, systems, and services. In addition to its library of standards, ASTM publishes several technical journals.
- The Institute of Electrical and Electronics Engineers, Inc. (IEEE) is a nonprofit professional organization that develops international standards for the advancement of technology, particularly in areas such as aerospace systems, telecommunications and information technology, biomedical engineering, and power generation. The standards are developed from research and technical committees comprised of technical experts.
- The International Standards Organization (ISO) is an international consortium that develops technical standards by consensus. The standards define specifications and criteria for the classification of materials, manufacture and supply of products, testing and analysis, terminology, and services. Although the ISO is not an enforcement agency, ISO 9000 certification is an international quality benchmark that verifies conformance to ISO standards.
- The National Fire Protection Association (NFPA) is an international nonprofit organization for fire, electrical, building, and public safety. The NFPA forms technical committees, adopts and publishes standards, provides training, and is a global leader in the prevention of fire and related hazards and building safety standards. The NFPA publishes standards such as the *National Fire Codes*, the *National Electric Code*, and the *Life Safety Code* through its technical committees.
- The Society of Automotive Engineers (SAE) is an international nonprofit organization that advances the engineering practices of mobility systems. Its technical committees publish specifications and guidance for materials, products, processes, and procedures for powered vehicles.
- The Society of Fire Protection Engineers (SFPE) is an nonprofit international organization that supports practitioners in fire protection engineering. Technical committees develop and publish guidelines and practices for fire protection designs and services.

Several of these nonprofit organizations develop, adopt, and publish industry standards. All use technical expertise from industry, government, academia, and other stakeholders in developing models and best practices for managing safety engineering work.

Approaches to Managing Safety Engineering Work

Safety engineering is managed and structured like other business plans within the context of an organization and its system-safety program. A business plan implements a strategy "to look ahead, allocate resources, focus on key points, and prepare for problems and opportunities" (Berry 2004). The foundation for a business plan includes a mission statement, goals, responsibilities, a decision process, a budget, and performance measurements. A system-safety plan is similar to a business plan in its overall structure and essential management elements. Effective management of safety engineering work requires similar planning, allocation of resources, human resource management, and performance measurements. In addition, management of safety engineering work addresses communications, which may be accomplished by using an established software program and writing reports.

Besides outlining the steps and key decision points in a given management model, the system-safety plan should identify the plan's interactions with other key functions. Some organizations use internal system-safety plans, but if they contract to provide a service for another organization, it is prudent to examine the interface points. Many other system or process management models use common approaches in assessment strategies, including a life-cycle framework, early integration, risk management, decision points, performance assessments, user/operator/stakeholder feedback, and continuous improvement.

One of the best practices in the management of safety engineering work is forming a cross-functional team, often referred to as a working group, which encourages collaboration and communication. A team approach allows the disciplines to design and evaluate a system, and communication is often facilitated with design reviews and periodic presentations.

A multidisciplinary team-based approach to life-cycle management creates synergism in assessing identified, unknown, and potential hazards that may have an adverse effect on people, property, and the environment. As stated by Belke (2000), "... various elements are generally not independent from one another; a significant interrelationship usually exists between each of them, and all of the elements need to work well together, or accidents can happen." A major benefit in using a multidisciplinary team is to identify a broader range of alternatives to eliminate or minimize hazards. Unintended consequences could result from mitigating hazards in one area that could create problems in another. For example, in the design of a new high-pressure research facility in the United Kingdom, a multidisciplinary team was formed to address both system development and operations. There were a series of modifications to both equipment and the facility. The working group was an essential part of the engineering design throughout the process. The group made it possible to identify alternative equipment configurations that met operational needs within a new facility (Philbin 2010).

Risk Management in a Multidisciplinary Environment

Risk management is a process that is used in several disciplines as an analytical method to evaluate requirements, costs, and risks, and to make decisions about the system. Often, each engineering discipline will use different models for risk assessment. It is beneficial to identify the various engineering disciplines that will be involved in the development of a system and use a common risk-assessment matrix upon which to evaluate hazards.

Research and case studies of the experiences of many organizations demonstrate that consideration and integration of safety engineering and other related disciplines facilitate hazard identification and corrective actions, which are especially important during early phases when corrective actions can be less complex and costly.

A best practice in managing safety engineering work is to be aware of other disciplines and of opportunities to collaborate or coordinate information with them. For example, environmental engineering, the application of science and engineering, studies systems and processes that mitigate harm to the environment. Environmental concerns considered early in the design can address systems and processes that may affect air, ground, water, hazardous waste, or human health.

Most systems will require human involvement in some aspect of operations, maintenance, or disposal, thus allowing a potential exposure to hazards, or the possibility of human error. The field of human factors engineering is involved with the control of hazards in a human–machine interface through design. Scientific and engineering methods are used to assess the potential for human error and equipment or system design hazards with the goal of designing jobs, machines, operations, environments, and work systems for compatibility with human capabilities and limitations. Many ergonomic concerns, such as musculoskeletal disorders (MSDs), high noise levels, or eye hazards, may be minimized by utilizing human factors engineers to assess the risks that may affect the user or operator of the system. For example, the seat dimensions, head and leg room, and placement and illumination of visual displays and operator controls in aircraft cockpits are designed to fit aviators. The goal of the system is to maximize the space and to consider the tasks and operations in the cockpit in order to minimize human error in both routine and emergency situations.

Collaboration with other organizations improves each organization's ability to identify potential hazards in a system. The Department of Defense (DOD), Federal Aviation Administration (FAA), and National Aeronautics and Space Administration (NASA) jointly developed a publication on engineering design criteria and guidelines entitled *Human Engineering Design Data Digest* (2000). This publication provides quantitative human engineering design criteria and guidelines for use during system, equipment, or facility design and assessment. In the management of safety engineering work, the collaboration and use of publications, databases, and other methods that help to identify hazards is a best practice that improves safety and efficiency.

Other engineering disciplines also enhance system design and efficiency. Every change in materials, equipment, process, or operation in a system's life cycle should be assessed by a safety engineer to identify potential hazards that may impact the system.

A Comprehensive Management Framework

The field of safety has emerged from a need to reduce accidents, minimize losses, prevent harm, and increase efficiency. In industrial safety, that need has eventually materialized into laws and regulations that standardize requirements for various hazardous working conditions. From that perspective, the federal government has specified the requirements and enforced the laws and regulations, thus building the foundation for industrial safety management. In contrast, system safety has evolved from responsible organizations that have a need to produce a safe system or product. Historically, organizations with the greatest need to produce safe systems have been developers in the aerospace, aviation, missile, and nuclear industries, primarily government and government contractors. The concept of system safety has grown and expanded to other industries, such as transportation; however, the structure and implementation of system safety is dependent on organizational requirements rather than law.

Generally, the framework for managing safety engineering work commonly includes the following elements:

1. Defining roles and responsibilities
2. Determining requirements and criteria
3. Allocating resources
4. Identifying a strategy and process to identify hazards and manage risk
5. Measuring performance
6. Facilitating communication, information exchange, and a feedback loop
7. Facilitating participation, involvement, and commitment to safety and health at all levels in the organization

There are numerous ways to manage safety engineering work. Organizational procedures vary on the management structure, safety engineering methodologies, risk categories, and other elements of system safety. Organizations such as the DOD, FAA, NASA, DOE, DOT, and others have similar system-safety programs and regulations that provide a comprehensive framework for managing safety engineering work for the systems they develop.

In private industry, organizations also establish the parameters for developing new systems, and, as with the federal government agencies, each within their scope of authority. MacCollum suggests using a similar methodology for the construction industry (MacCollum 2008, 26): "The application of system safety principles has been largely limited to the aerospace and electronic community. These principles need to be adopted by forward-thinking design and build contractors."

Although each organization varies in its mission, collaboration and communication within industries serve to help and encourage organizations to use and improve both management and safety engineering work that results in systems that cost less and are safer to build, operate, and maintain. In the construction industry, a long-standing organization, the National Institute of Building Sciences (NIBS), was established in 1974 to serve as an interface between government and the private sector. Their mission is to serve the nation by advancing knowledge in the building sciences and technology to enable organizations to identify and resolve issues for the construction of safe, affordable structures for housing, commerce, and industry (NIBS 2011). Through the various committees comprised of public and private industry representatives, the organization develops and publishes consensus standards that facilitate communication and an integrated approach to building safe structures.

However, although such industry consensus standards promote sound concepts and standards, each organization must determine the structure for management and implementation within their organizations. As Vincoli stated, ". . . no matter how exact and comprehensive a design or operating safety program is considered to be, the proper management of that system is still one of the most important elements of success" (Vincoli 1997, 5).

Organizations that achieve the greatest success in developing and maintaining safe systems recognize that the management framework and safety engineering work are two components of a comprehensive system. The third intangible component that is highly influential in how the work is carried out, and the decisions made concerning the system, is the organizational culture.

Conclusion

This chapter discusses the best practices in managing safety engineering work. There is a difference between the definition of system safety and safety engineering. The subtle distinction is that system safety is defined as the management system, and safety engineering is the application of safety and engineering within the management framework.

Best practices in the management of safety engineering work are based on sound business principles and administration. There are many ways to manage safety engineering work, using a variety of sources for information, including regulatory agencies, consensus standards, and professional organizations and publications, all of which are intended to facilitate an environment of continuous learning and improvement. There are three fundamental components in successfully managing safety engineering work that will reap the greatest benefit to the organization and the system: (1) organizational culture, (2) management framework, and (3) safety engineering practices. The three components are intertwined and interdependent.

At an organizational level, a safety culture fosters a dynamic learning environment committed to continuous improvement through the integration of safety management into all facets of work planning and execution. The comprehensive framework is a structured model that integrates safety as part of the design and throughout other phases in the life cycle of a product, process, or facility. The primary function of system safety is to identify and control hazards in each phase of the life cycle, from concept through decommissioning and disposal.

A management system provides a process for planning, organizing, budgeting, and aligning resources within time and cost constraints. Within these major functions, there are various elements that define the management process and tasks, including decision making, implementation, and evaluation.

This chapter shows how safety engineering work without a supportive organizational culture can influence risk decisions concerning cost and schedule. Similarly, the best organizational culture or management system without the technical expertise of safety engineering methodology can contribute to the failure in any system. As stated by Main: "The best approach for a particular company is the method that works best in the organizational culture and design process" (Main 2004, 423).

References

American Petroleum Institute (API). *Recommended Practice for Development of a Safety and Environmental Management Program for Offshore Operations and Facilities* (accessed February 1, 2011). www.api.org/Standards/epstandards/upload/75_E3.pdf

――――. *Specification for Quality Programs for the Petroleum, Petrochemical and Natural Gas Industry* (accessed February 3, 2011). www.api.org/Standards/faq/upload/valueofstandards.pdf

Belke, James C. 2000. *Chemical Accident Risks in U.S. Industry: A Preliminary Analysis of Accident Risk Data from U.S. Hazardous Chemical Facilities* (accessed February 3, 2011). www.toxic.dead-planet.net/pdfs/stockholmpaper.pdf

Berry, Tim. 2004. *Hurdle: The Book on Business Planning.* Eugene, OR: Palo Alto Software, Inc.

Brauer, R. 1990. *Safety and Health for Engineers.* New York: Van Nostrand Reinhold.

Clemens, P., and R. Simmons. 2000. *System Safety Scrapbook* (accessed February 3, 2011). www.fault-tree.net/papers/clemens-safety-scrapbook.pdf

Department of Defense (DOD). 2000. *Human Engineering Design Data Digest* (accessed February 7, 2011). www.dtic.mil/cgi-bin/GetTRDoc?AD=ADA467401&Location=U2&doc=GetTRDoc.pdf

_____. 2005. Military Standard 882E, *Standard Practice for System Safety* (accessed January 17, 2011). www.system-safety.org/Documents/MIL-STD-882E-Feb05.doc

Food and Drug Administration (FDA). *Facts About Current Good Manufacturing Practices (CGMP)* (accessed February 7, 2011). www.fda.gov/Drugs/DevelopmentApprovalProcess/Manufacturing/ucm169105.htm

Haimes, Yacov Y. 2009. *Risk Modeling, Assessment, and Management.* Hoboken, NJ: John Wiley & Sons.

Leveson, N. 2004. *A New Accident Model for Engineering Safer Systems* (accessed February 9, 2011). sunnyday.mit.edu/accidents/safetyscience-single.pdf

_____. 2009. *Engineering a Safer World: System Safety for the 21st Century* (accessed February 9, 2011). sunnyday.mit.edu/book2.pdf

Leveson, N., and J. Cutcher-Gershenfeld. 2004. *What System Safety Engineering Can Learn from the Columbia Accident* (accessed February 10, 2011). sunnyday.mit.edu/papers/issc04-final.pdf

Leveson, N., J. Cutcher-Gershenfeld, B. Barrett, A. Brown, J. Carroll, N. Dulac, L. Fraile, and K. Marais. 2004. "Effectively Addressing NASA's Organizational and Safety Culture: Insights from Systems Safety and Engineering Systems." Paper presented at the Engineering Systems Division Symposium, Massachusetts Institute of Technology, Cambridge, MA, March 29–31, 2004 (accessed March 17, 2011). sunnyday.mit.edu/papers/esd2-columbia.doc

MacCullom, D., and R. Davis. 2008. "Engineering Principles for Safer Design." Presented at the Professional Development Conference of the American Society of Safety Engineering June 9–12, 2008, Las Vegas, NV.

Main, B. 2004. *Risk Assessment: Basics and Benchmarks.* Ann Arbor, MI: Design Safety Engineering, Inc.

Marshall, G. 2000. *Safety Engineering.* 3d ed. Des Plaines, IL: American Society of Safety Engineers.

National Aeronautic and Space Administration (NASA). 1999. DHB-S-001, *System Safety Handbook.* hnd.usacc.army.mil/safety/RefDocs/FASS/NASA%205Systems%20Safety.pdf

_____. 2008. NPR 8715.3, *General Safety Program Requirements* (retrieved January 3, 2011). www.hq.nasa.fov/office/code/doctree/87153.htm

_____. 2007. SP-2007-6105 Rev. 1, *System Engineering Handbook.* education.ksc.nasa.gov/esmdspacegrant/Document/NASA%20SP-2007-6015%20rev%201%20Final%2031Dec2007.pdf

National Institute of Building Sciences (NIBS). 2011. *Whole Building Design Guide* (accessed April 12, 2011). www.nibs.org

Nuclear Regulatory Commission (NRC). 2005. *Safety Program and Integrated Safety Analysis* (accessed February 7, 2011). www.nrc.gov/reading-rm/doc-collections/cfr/part070/part070-0062.html

Philbin, Simon P. 2010. "Developing an Integrated Approach to System Safety Engineering." *Engineering Management Journal* 22.2(2010): 56–67.

Roland, Harold E., and B. Moriarty. 1990. *System Safety Engineering and Management.* 2d ed. Hoboken, NJ: John Wiley & Sons, Inc.

Schein, E. H. 2004. *Organizational Culture and Leadership.* 3d ed. San Francisco, CA: John Wiley & Sons, Inc.

Stephans, Richard A. 2004. *System Safety for the 21st Century: The Updated and Revised Edition of System Safety 2000.* Hoboken, NJ: Wiley-Interscience.

Turner, R. 2003. "Seven Pitfalls to Avoid in the Hunt for Best Practices." *IEEE Software* 20(1):67–69.

Vincoli, J. 1997. *Basic Guide to System Safety.* New York: John Wiley & Sons.

Wasson, C. 2006. *System Analysis, Design, and Development: Concepts, Principles, and Practices.* Hoboken, NJ: John Wiley & Sons, Inc.

Appendix: Selected References, Publications, and Organizations

The following references are a few of the organizations and publications where the reader can find best practices in systems and engineering.

American Chemistry Council (ACC). *Model of Responsible Care*. www.americanchemistry.com

American Institute of Chemical Engineers (AIChE). *Technical Management of Chemical Process Safety*. www.knovel.com/knovel2/Publisher.jsp?PublisherID=101253

American National Standards Institute (ANSI). www.ansi.org

American Petroleum Institute (API). www.api.org

American Society of Civil Engineers (ASCE). *Journal of Construction Engineering and Management*. www.asce.org/asce.cfm

American Society of Engineering Education (ASEE). www.asee.org

American Society of Heating, Refrigeration, and Air Conditioning Engineers (ASHRAE). www.ashrae.org

American Society of Mechanical Engineers (ASME). www.asme.org

American Society of Mechanical Engineers, Safety Engineering and Risk Analysis Division (SERAD). www.asme.org/divisions/serad

American Society of Safety Engineers (ASSE). www.asse.org

Architectural Engineering Institute (AEI). www.content.aeinstitute.org/intro.html

Department of Defense (DOD). *Defense Acquisition Guidebook*, "Systems Engineering." www.defenselink.mil/pubs/archive/html

Department of Defense (DOD). Ergonomics Working Group, *Best Practices*. www.ergoworkinggroup.org/ewgweb/IndexFrames/index3.htm

Department of Defense (DOD), Human Factors Engineering Technical Advisory Group. *Human Engineering Design Data Digest*. www.hfetag.dtic.mil/docs/pocket_guide.doc

Department of Energy (DOE). DOE P 450.4, :"Safety Management System Policy." www.directives.doe.gov/pdfs/doe/doetext/neword/450/p4504.html

Electric Power Research Institute (EPRI). www.my.epri.com

Federal Aviation Administration (FAA). *System Safety Handbook*. www.faa.gov/library/manuals/aviation/risk_management/ss_handbook

Food and Drug Administration, Center for Drug Evaluation and Research. *Current Good Manufacturing Practices (CGMP) Regulations: Division of Manufacturing and Product Quality*. www.fda.gov/cder/dmpq

Food and Drug Administration (FDA), Center for Drug Evaluation and Research. *Good Laboratory Practices, A Risk-Based Approach to Pharmaceutical Current Good Manufacturing Practices (CGMP) for the 21st Century*. www.fda.gov/cder/gmp

Food and Drug Administration (FDA). *Good Laboratory Practices (GLP)*. www.fda.gov/ora/compliance_ref/bimo/glp/default.htm

Human Factors: The Journal of the Human Factors and Ergonomics Society. www.hfes.org/web/AboutHFES/history.html

Institute of Electrical and Electronics Engineers (IEEE). www.ieee.org

Institute of Physics, Nanotechnology. www.nanotechweb.org/

International Ergonomics Association. 2000. *The Discipline of Ergonomics*. www.iea.cc/

International Standards Organization (ISO). www.iso.org

Johnson, Paul W. *ePORT, NASA's Computer Database Program for System Safety Risk Management Oversight (Electronic Project Online Risk Tool)*. National Aeronautics and Space Administration. ntrs.nasa.gov/archive/nasa/casi.ntrs.nasa.gov/20080030091_2008022873.pdf

Joint Commission Accreditation of Healthcare Organizations (JCAHO). www.jcaho.org

Journal of Architectural Engineering. www.pubs.asce.org/journals

Mechanical Engineering Magazine Online. www.memagazine.org

National Aeronautics and Space Administration (NASA). *Systems Engineering Handbook*. www.ntrs.nasa.gov/archive/nasa/casi.ntrs.nasa.gov/19960002194_1996102194.pdf

National Fire Protection Association (NFPA). www.nfpa.org

National Nanotechnology Initiative (NNI). www.nano.gov

Society of Automotive Engineers (SAE). www.sae.org

System Safety Society. www.system-safety.org

U.S. Department of Energy (DOE), Nuclear Criticality Safety Program. www.ncsc.llnl.gov

INDEX

A

ABA. *See:* American Bankers Association (ABA)
ABET. *See:* Accreditation Board of Engineering and Technology (ABET)
Accidents. *See:* Hazards *and* Incident investigations
Accreditation Board of Engineering and Technology (ABET), 65–66
ADA. *See:* Americans with Disabilities Act (ADA)
AHJ. *See:* Authority having jurisdiction (AHJ)
AIHA. *See:* American Industrial Hygiene Association (AIHA)
Alliance programs, 6
Allocated overhead charges, 132
Allowances, for fatigue and delays, 95–96
American Bankers Association (ABA), 121
American Industrial Hygiene Association (AIHA), 52, 55
American National Standards Institute (ANSI)
    A10 Standard, 154, 158–159
    consensus standards, 3, 5, 180
    risk-assessment methods, 30
    Z10 Standard, 17–18, 24–26, 30, 34–35, 42–43, 46–47, 107, 180
    Z15 Standard, 59–64
American Petroleum Institute (API), 180
American Society for Testing and Materials (ASTM), 181
American Society of Mechanical Engineers (ASME), consensus standards, 4, 5
American Society of Safety Engineers (ASSE)
    education, safety professional, 49–53
    legal perspective on ANSI Z10, 42
    resources, 180–181
    Safety Body of Knowledge (BoK), 55
Americans with Disabilities Act (ADA), 68–69
Annual cost analysis, 146
ANSI. *See:* American National Standards Institute (ANSI)
ANSI Z15.1Standard: A Tool for Preventing Motor Vehicle Injuries and Minimizing Legal Liability (Abrams), 59–64
API. *See:* American Petroleum Institute (API)
Apollo process, 39
Arithmetic gradient, 139
Asbestos, 110
ASME. *See:* American Society of Mechanical Engineers (ASME)
ASSE. *See:* American Society of Safety Engineers (ASSE)
Assessment. *See:* Benchmarking *and* Performance
ASTM. *See:* American Society for Testing and Materials (ASTM)
ATM. *See:* Automatic teller machine (ATM)
Authority having jurisdiction (AHJ), 2–3, 5, 12
Automatic teller machine (ATM), 121

B

BCSP *See:* Board of Certified Safety Professionals (BCSP)
Belke, James, 182
Benchmarking, 94, 151–170. *See also:* Performance
    culture and safety, 154
    cumulative frequency distributions, 165
    effectiveness, 151, 154–157
    ISO standards, 169–170
    overview, 151
    performance measure criteria, 167
    process and steps, 157–158
    quality control, 158–159
    safety management system, 152–154
    statistical process control, 159–165
    valid measurements, 155–157
Benefit-cost ratio, 144–145
Best management practices. *See:* Management *and* specific fields
Binomial distributions, 166
Bird, F. E., 130
Blair, Earl, 51, 54
Blanchard, Kenneth, 81–82
BLS. *See:* Bureau of Labor Statistics (BLS)
Board of Certified Safety Professionals (BCSP), 51, 55
BP *See:* British Petroleum (BP)
Brauer, Roger, 66, 174
Break-even analysis, 144
British Petroleum (BP), 102, 110
    case study, 103
    Texas City refinery, 179
Budgeting, 66, 76, 146–148. *See also:* Cost analysis, Costs, *and* Economic analysis
    overview, 129–130
    process, 146–147
    safety management, 76–79
    uncertainty, 147–148
Bureau of Labor Statistics (BLS), 40, 144

C

CAA. *See:* Clean Air Act (CAA)
Canadian Registered Safety Professional (CRSP), 119
Canadian Workplace Hazardous Materials Information Systems (WHMIS), 115
Capital assets pricing model (CAPM), 146
The Career Guide to the Safety Profession (ASSE), 50
Cash flow diagrams (CFDs), 136–137
CDC. *See:* Centers for Disease Control and Prevention (CDC)
Centers for Disease Control and Prevention (CDC), 122
CEQ. *See:* Council on Environmental Quality (CEQ)
CERCLA. *See:* Comprehensive Environmental Response, Compensation, and Liability Act (CERCLA)
Certifications, professional, 51–53
Certified Industrial Hygienist (CIH), 52, 119
Certified Safety Professional (CSP), 51, 119
CESB. *See:* Council of Engineering and Scientific Specialty Board (CESB)
CFDs. *See:* Cash flow diagrams (CFDs)
CFR. *See:* Code of Federal Regulations (CFR)
CGA. *See:* Compressed Gas Association (CGA)
Change analysis, 28, 29
Chemical Safety Board (CSB), 179
Chemical Safety Information, Site Security and Fuels Regulatory Relief Act (CSISSFRRA), 1999, 11
CIH. *See:* Certified Industrial Hygienist (CIH)
CIP *See:* Continuous improvement process (CIP)
Classroom training. *See:* Training
Clean Air Act (CAA), 1963, safety engineering management, 8
Clean Water Act (CWA), 1977, safety engineering management, 9
Clemens, P., 174
Code of Federal Regulations (CFR), 2, 5, 179–180
Communication, global, 119–120
Community Right to Know Law. *See:* Emergency Planning and Community Right to Know Act (EPCRA), 1986
Competency, safety professional, 54–55
Compliance, safety engineering, 18–21, 24, 26, 90
Compound interest, 139–140
Comprehensive Environmental Response, Compensation, and Liability Act (CERCLA), 1980, 9–10
Comprehensive surveys, 24, 29
Compressed Gas Association (CGA), 3, 5
Conflict, managing, 84–85
Conformity assessment, 113–114
Construction, safety engineering, 66–68, 70, 87, 96, 135, 183
Continuous improvement process (CIP), 17, 158–159, 168
Control chart, 160, 163–164
Cooperative programs, 5–7
Corporate social responsibility (CSR), 110

Cost analysis, 129–148. *See also:* Budgeting *and* Economic analysis
    cash flow, 136–140
    direct costs, 133–134
    expected value, 142
    indirect costs and benefits, 131–136
    intangible costs and benefits, 132–133
    investment analysis, 142–146
    life-cycle costs, 130–131
    literature on safety costs, 133–136
    overview, 129–130
    time value of money, 137–139
Costs
    direct, 133–134
    intangible, 132–133
    life-cycle, 130–131
    literature on safety costs, 133–134
    occupational injuries, 15–16, 21, 34
    opportunity, 137
    overhead charges, 102
    project safety budgets, 75–79
    sunk, 147
    system safety framework, 179
Council of Engineering and Scientific Specialty Board (CESB), 53
Council on Environmental Quality (CEQ) report, 7–8
CoVan, James, 65–66
CRSP. *See:* Canadian Registered Safety Professional (CRSP)
CSB. *See:* Chemical Safety Board (CSB)
CSISSFRRA. *See:* Chemical Safety Information, Site Security and Fuels Regulatory Relief Act (CSISSFRRA)
CSP. *See:* Certified Safety Professional (CSP)
CSR. *See:* Corporate social responsibility (CSR)
CTDs. *See:* Cumulative trauma disorders (CTDs)
Culture, organizational
    evaluation form, 44
    global strategies, 110–111
    incentives and motivation, 85–87
    leadership, 177
    safety engineering management, 25–27, 41, 43, 45–46, 177–179
Cumulative frequency distributions, 165
Cumulative trauma disorders (CTDs), 155–156, 165
Current indicators, safety effectiveness, 156
Customers, objectives, 68
CWA. *See:* Clean Water Act (CWA)

D

Defense Nuclear Facilities Safety Board (DNFSB), 3
DeLeo, William, 54–55
Deming, W Edwards, 158–159
Depreciation, 132
DeSiervo, Robert, 47
Direct costs, 133–134

Discount rate, 137
DNFSB. *See:* Defense Nuclear Facilities Safety Board (DNFSB)
DoDI. *See:* U.S. Department of Defense Instruction (DoDI)
DOE. *See:* U.S. Department of Energy (DOE)
DOL. *See:* U.S. Department of Labor (DOL)
DOT. *See:* U.S. Department of Transportation (DOT)
Dummy activities, 71

E
Economic analysis. *See also:* Cost analysis
    project safety budgets, 75–79
    safety management, 73–74
Economic Control of Quality of Manufactured Product (Shewhart), 160
Education, safety professional, 49–55. *See also:* Training
Effective interest rate, 139
Effectiveness. *See:* Benchmarking *and* Performance
EIS. *See:* Environmental Impact Statement (EIS)
Eisner, H., 80–81, 83
Emergency contacts, 120–121
Emergency Planning and Community Right to Know Act (EPCRA), 1986, 10
Employees. *See also:* Training *and* Culture, organizational
    performance, 87, 94–95
    safety and health standards, 72, 74–75
    safety management, role in, 19, 27–29, 36, 45–46, 152
    teams, 82–86
    Voluntary Protection Program, 167–168
Employers. *See also:* Leadership *and* Management
    culture and safety, 26–28, 37, 41–46
    General Duty Clause (OSHA) obligations, 18–21
    integrated workplace safety approach, 106–109
    training programs, 36
    Voluntary Protection Program, 167–168
Endangered Species Act (ESA), 1973, 8
Engineering. *See also:* Globalization *and* Safety engineering
    defined, 65–66
    economics, 76–79
    human factors, 182–183
Environmental Impact Statement (EIS), 7
Environmental Protection Agency (EPA)
    Office of Pollution Prevention and Toxics (OPPT), 9
    program-management regulations, 7–11
    safety engineering management, 180
EPCRA. *See:* Emergency Planning and Community Right to Know Act (EPCRA), 1986
EPZ. *See:* Export-processing zone (EPZ)
Equivalency, 137–139
Equivalent annual benefit and cost analysis. *See:* Annual cost analysis
ESA. *See:* Endangered Species Act (ESA)
Expected value, 142
Experience, safety professional, 47–49, 52–53
Export/import, global, 102–104
Export-processing zone (EPZ), 102

F
5 Whys, 38
FAA. *See:* Federal Aviation Administration (FAA)
Failure modes and effects analysis (FMEA), 31, 178
Fair Labor Association (FLA), 104
Fatal Facts, (OSHA), 49
Fault tree analysis (FTA), 31, 178
FDA. *See:* Food and Drug Administration (FDA)
Federal Aviation Administration (FAA), 180, 182
Federal Food, Drug, and Cosmetic Act (FFDCA), 10
Federal Motor Carrier Safety Administration (FMCSA), 62
Federal Motor Vehicle Safety Standards (FMVSS), 180
Federal Register, 2–3
Feedback controls, 96
FFDCA. *See:* Federal Food, Drug, and Cosmetic Act (FFDCA)
Finances. *See:* Budgeting, Cost analysis, Economic analysis, *and* Investment
Fixed costs, 132
FLA. *See:* Fair Labor Association (FLA)
FMCSA. *See:* Federal Motor Carrier Safety Administration (FMCSA)
FMEA. *See:* Failure modes and effects analysis (FMEA)
FMVSS. *See:* Federal Motor Vehicle Safety Standards (FMVSS)
Food and Drug Administration (FDA), 10, 112, 180
Freivalds, A., 86
FTA. *See:* Fault tree analysis (FTA)
Future value (FV), 139

G
Gantt charting, 73–75
GBCI. *See:* Green Building Certification Institute (GBCI)
General Duty Clause (OSHA), 18–21
Geometric gradient, 139
Global Alliance for Workers and Communities, 104
Global Reporting Initiative (GRI), 105–106
Global safety and health risk profile (GRP), 109–110
Globalization, 101–123
    business perspective, 102
    communication, 119–120
    competency and expertise, 118–119
    conformity and compliance, 113–115
    health and risk profile, 109–110
    indicators, safety and health performance, 116–118
    and OSHA, 111–112

Globalization (*cont.*)
    overview, 101–102
    reputational risk, 102–104
    resources, 116, 118–119
    safety and health management systems, 112–115
    strategy and implementation, 110–111
    travel and personal risks, 120–122
    workplace safety challenges, 104–109
GRI. *See:* Global Reporting Initiative (GRI)
Groupthink, 83
GRP. *See:* Global safety and health risk profile (GRP)

H
Habits of mind, 86
Hazard, analysis, 16–18, 29–33
Hazard and operability analysis (HAZOP), 31
Hazard prevention, safety engineering management, 30–36
Hazard ranking system (HRS), 10
Hazards
    assessment guide, 30–33
    defined, 15–16
    General Duty Clause (OSHA) obligations, 18–21
    incident investigations, 37–39
    work-site analysis, 16–18, 28–30, 153
HAZOP. *See:* Hazard and operability analysis (HAZOP)
Health management, global challenges, 104–109. *See also: safety entries*
Hersey, P., 81–82
Hidden costs, 132–133
Hierarchy of needs theory, 85
HRS. *See:* Hazard ranking system (HRS)
Human Engineering Design Data Digest (DOD, FAA, NASA), 183
Human factors engineering, 182–183
Human resources, 69, 87–89

I
IAEA. *See:* International Atomic Energy Agency (IAEA)
IEEE. *See:* Institute of Electrical and Electronics Engineers, Inc.
Import/export, global, 102
Incident investigations
    causes, 37–38
    vs intervention application rate, 88–90
Income tax benefits, 131–132
Increasing Productivity and Profit through Health and Safety (Oxenbaugh, Marlow, & Oxenbaugh), 130
Indicators, safety and health performance, 116–118
Indirect costs, 133–136
Industrial Health and Safety Act (Korea), 115
Inflation, 140–141
Inspections, safety and health, 32, 96–97
Institute for Work and Health, Ontario, Canada, 45

Institute of Electrical and Electronics Engineers, Inc. (IEEE), 4, 180–181
Interest rates, 139–140
Internal Control: Guidance for Directors on the Combined Code (Turnbull Report), 105
Internal rate of return (IRR), 145
International Atomic Energy Agency (IAEA), 178
International issues. *See:* Globalization
International Organization for Standardization (ISO)
    9000 Standard, 7, 113, 159, 167
    9001:2008 Standard, 158, 169, 180
    9004:2009 Standard, 169
    14000 Standard, 7, 113, 158, 167–170
    14001:2004 Standard, 169, 170
    14004:2004 Standard, 170
    26000:2010 Standard, *Guidance on Social Responsibility*, 106
    accreditation, 53
    family of standards, 169–170
    Plan-Do-Check-Act management, 158–159
    resources, 181
    safety engineering management, 4, 7
Introduction to Engineering (Wright), 65
Introduction to Safety Engineering (Gloss & Wardle), 66
Investigations. *See:* Incident investigations
Investment. *See also:* Cost analysis *and* Economic analysis
    analysis, 142–146
    indirect costs and benefits, 131–132
IRR. *See:* Internal rate of return (IRR)
Ishikawa, Kaoru, 38
ISO. *See:* International Organization for Standardization (ISO)

J
Job hazard analysis (JHA), 152–153
Job safety analysis (JSA), 152

K
Kepner-Tregoe (KT), 39

L
Leadership, 25–26, 28–29, 41–43. *See also:* Culture, organizational *and* Management
    attributes, 79–80
    commitment and globalization, 105–109
    competency and expertise, 118–120
    conflict, 84–85
    integration, 106–109
    motivation and incentives, 85–87
    safety management, 152
    self-evaluation, 80–81
    situational, 81–82
    of teams, 82–84

Leading indicators, safety effectiveness, 156
Learning curves, 90–92
LEPCs. *See:* Local emergency planning committees (LEPCs)
Leveson, N., 174, 178
Liberty Mutual Insurance Company, 60
Life-cycle costs, 130–131
Life Cycle Management Manual (U.S. Patent and Trademark Office), 180
Life Safety Code, 181
Local emergency planning committees (LEPCs), 10
L'Oreal, 115–116
Lost-time case rate (LTCR), 116–117
Lost-workday injuries and illnesses (LWDII), 39–40

M

MacCollum, D., 183
MACRS. *See:* Modified Accelerated Cost Recovery System (MACRS)
Main, Bruce, 178
Maine Employers' Mutual Insurance Company (MEMIC), 44–45
Management. *See also:* Culture, organizational; Leadership; *and* Safety engineering management
    approaches, 67
    attributes, 79–80
    competency and expertise, 118–119
    corporate social responsibility, 110
    culture and safety, 26–27, 43–46, 111–113, 154
    health management system, 16–18, 25–26
    safety practitioners, 47–55
    system safety engineering, 173–177
Management of Environment, Safety, and Health (MESH) program (Nike), 104
Manuele, Fred A., 16–17, 31, 33, 37, 42–43, 46–47
Maslow, Abraham, 85
Maximum containment level (MCL), 8–9
Maximum containment-level goal (MCLG), 9
Medical safety, 122
MEMIC. *See:* Maine Employers' Mutual Insurance Company (MEMIC)
Merit Program, 168
MESH. *See:* Management of Environment, Safety, and Health (MESH)
Methods Standards and Work Design (Niebel & Freivalds), 86
Military Standard 882E, 175–176
Mine Safety and Health Administration (MSHA), 3, 5
Modified Accelerated Cost Recovery System (MACRS), 78
Motivation. *See also:* Culture, organizational
    defined, 85
    of workers, 85–87
MSHA. *See:* Mine Safety and Health Administration (MSHA)

N

NAAQS. *See:* National Ambient Air Quality Standards (NAAQS)
National Advisory Committee on Safety and Health (NACOSH), 62
National Aeronautics and Space Administration (NASA), 175, 177–178, 182
National Ambient Air Quality Standards (NAAQS), 8
National Archives and Records Administration (NARA), 2
National Bureau of Statistics (China), 102
National Commission on Certifying Agencies (NCCA), 53
National Electric Code, 181
National Environmental Policy Act (NEPA), 1969, 7–8
National Fire Codes, 181
National Fire Protection Association (NFPA), 3, 5, 12, 181
National Highway Traffic Safety Administration (NHTSA), 180
National Institute of Building Sciences (NIBS), 176, 183
National Safety Council (NSC), 37, 177
NCCA. *See:* National Commission on Certifying Agencies (NCCA)
Needs assessment, 118
NEPA. *See:* National Environmental Policy Act (NEPA)
Net present value (NPV) *or* net present worth (NPW), 77–79, 139, 145–146
NFPA. *See:* National Fire Protection Association (NFPA)
NHTSA. *See:* National Highway Traffic Safety Administration (NHTSA)
NIBS. *See:* National Institute of Building Sciences (NIBS)
Niebel, B., 86
Nike, Inc., 102, 104, 110
Nodes, 71–72
Nominal rate, 139
Np control charts, 162–163, 165
NPV. *See:* Net Present Value (NPV)
NSC. *See:* National Safety Council (NSC)
Nuclear Regulatory Commission (NRC), 180

O

Objectives, safety engineering management, 67–68
Occupational Health and Safety Assessment Series (OHSAS)
    18001:2007, *Occupational Health and Safety (OHS) Management Systems Specification*, 168–169
    18002, *Guidelines for the Implementation of OHSAS 18001*, 169
Occupational Health and Safety Management System (OHSMS), 24–26, 46–47, 113
Occupational injury statistics, 37
Occupational Safety and Health Administration (OSHA)
    accident rates, 39–41
    Challenge Program, 7
    consultation program history, 21–22
    cooperative programs, 5–7
    cost savings, employees, 15–16

Occupational Safety and Health Administration (OSHA) (*cont.*)
    executive orders, 2
    Fatal Facts, 49
    General Duty Clause, 18–21
    globalization, 111–112
    health management system, 16–18
    health program management guidelines, 1989, 23–24
    management leadership, 25–29
    motor-vehicle crash statistics, 59
    program evaluation profile (PEP), 36
    program-management regulations, 4–5
    role in safety management, 18–24
    Safety and Health Achievement Recognition Program (SHARP), 5–6, 40, 41
    Safety and Health Program Assessment Worksheet, 167
    Safety and Health Regulations for Construction (29 CFR 1926), 5
    $afety Pays program, 21, 136
    Standards for General Industry (29 CFR 1910), 5
    state regulations, 3
    training, 36
    Voluntary Protection Program (VPP), 5–7, 22, 116, 167–168
    work-site analysis, 28–30
Occupational Safety and Health Act (OSH Act) 1, 4–5, 31, 62, 167–168, 179
Occupational safety and health management system cycle, 154
OHSAS. *See:* Occupational Health and Safety Assessment Series (OHSAS)
Oil Pollution Act (OPA), 1990, 11
OPPT. *See:* Environmental Protection Agency, Office of Pollution Prevention and Toxics (OPPT)
Oregon OSHA (OR-OSHA), 42–43
Organizational culture. *See:* Culture, organizational
Organizational management, safety engineering, 176–177, 183–184
OSHA. *See:* Occupational Safety and Health Administration (OSHA)
OSH Act. *See:* Occupational Safety and Health Act (OSH Act)
OSHA Challenge Program, 7
OSHA Strategic Partnership Program (OSPP), 6

P

P control charts, 162, 164
Pareto Principle and charts, 166–167
Payback period, 142–144
PDCA. *See:* Plan-Do-Check-Act (PDCA) cycle
PEP. *See:* Program evaluation profile (PEP)
Performance. *See also:* Benchmarking
    criteria, 167
    effectiveness, safety, 151–154
    employee, 94–95
    ISO standards, 169–170
    overview, 151
    valid measurements, 155–157
Personal protective equipment (PPE), 37, 35, 68, 135
Personal security, 120–122
PERT. *See:* Program Evaluation and Review Technique (PERT)
Petersen, Dan, 11, 22, 26, 29, 34, 39, 41–42
PHA. *See:* Preliminary hazard analysis, (PHA)
Plan-Do-Check-Act (PDCA) cycle, 17, 146–147, 154, 158–159
Planning. *See also:* Emergency preparedness
    Gantt charting, 73–75
    Program Evaluation and Review Technique (PERT), 70–73
Poisson distribution, 93, 165–166
Pollution Prevention Act (PPA), 1990, 11
Post-action controls, 96
POTW. *See:* Publicly owned treatment works (POTW)
PPA. *See:* Pollution Prevention Act (PPA)
PPE. *See:* Personal protective equipment (PPE)
Preliminary hazard analysis, (PHA), 31
Presentation, results, 97–98
Present value (PV), 139
Prevention. *See:* Hazard prevention
Process control, 159–160
Process safety information (PSI) document, 74
Productivity, 134, 159–165
Profit, defined, 129
Profitability index. *See:* Benefit-cost ratio
Program Evaluation and Review Technique (PERT), 70–73
Program evaluation profile (PEP), 36
Projects, safety engineering
    closeout, 96–97
    completion, 96–98
    defined, 67
    objectives, 67–68
    requirements, 68
    results, presentation of, 97–98
Protective equipment. *See:* Personal protective equipment (PPE)
PSI. *See:* Process safety information (PSI) document
Public water system (PSWS), 9
Publicly owned treatment works (POTW), 9
PV. *See:* Present value (PV)

Q

Quality control, 158–159, 169

R

RCRA. *See:* Resource Conservation and Recovery Act (RCRA), 1976

Index

Records, safety engineering management, 33
Registered Safety Professional (RSP), Safety Institute of Australia, 119
Regulations, safety engineering, 2–12, 68–69. *See also: specific topics*
Reporting of Injuries, Diseases and Dangerous Occurrences Regulations (RIDDOR), 115, 117
Resource Conservation and Recovery Act (RCRA), 1976, purpose of, 9
Return on investment (ROI), 42, 76
RIDDOR. *See:* Reporting of Injuries, Diseases and Dangerous Occurrences Regulations (RIDDOR)
Risk
    assessment, 30–36, 93, 174–176, 182
    defining safety management roles, 46
    global perspective, 109–110
    reputational, 102–104
    travel and personal, 120–122
Risk management plan (RMP), 11
ROI. *See:* Return on investment (ROI)
RSP. *See:* Registered Safety Professional (RSP)

S

SAE. *See:* Society of Automotive Engineers (SAE)
Safe Drinking Water Act (SDWA), 1974, 8
Safety and the Bottom Line (Bird), 130
Safety and Health Achievement Recognition Program (SHARP), 5–6, 40, 41
Safety and Health for Engineers (Brauer), 66
Safety and Health Regulations for Construction, 5
Safety Body of Knowledge (BoK), 55
Safety Engineering (CoVan), 65–66
Safety engineering, 1–12
    best management practices, 4, 173–184
    Code of Federal Regulations (CFR), 2
    consensus standards, 3–4
    cooperative programs, 5–7
    described, 1–2, 11–12, 65–67, 173–174
    EPA program-management regulations, 7–11
    executive orders, 2
    ISO standards, 4, 7
    OSHA and, 4–5
    overview, 1–2
    state and local regulations, 3
Safety engineering best practices, 173–184
    federal regulations, 179–180
    human factors, 182–183
    management framework, 176–177, 183–184
    organizational leadership, 177
    overview, 173–174, 183–184
    planning, resources, and costs, 179
    professional resources, 180–181
    system safety framework, 173–177
Safety engineering management, 15–55, 65–99. *See also:* Globalization *and* Management
    approaches, 67, 181–182
    benchmarking and performance, 157–158
    budgeting, 75–79
    completion, 96–98
    controls hierarchy, 34–35
    culture and safety, 26–27
    defined, 15–16
    effectiveness, 22–25, 46–47
    employee involvement, 27–29
    expected value of hiring, 142, 144
    hazard prevention, 30–36
    health management system, 16–18
    improvement, 17, 46–47
    incident investigations, 37–39
    leadership, 25–26, 28–29, 41–43, 79–87
    multidisciplinary risk assessment, 182–183
    OSHA's role in, 16–24
    plan, project, 67–69
    professionals in, 47–55
    program model, 87
    roles for, 46–47
    scheduling, 70–75
    work-site analysis, 28–30
Safety Culture Profile Survey (Maine Employers' Mutual Insurance Company), 45
$afety Pays program (OSHA), 21, 136
Safety Professional Truth in Advertising Act, 52
Safety professionals
    job description, 47–48
    salary, 49, 51–52
SARA. *See:* Superfund Amendments and Reauthorization Act (SARA)
Sarbanes-Oxley Act, 2002, 64, 105–106
Savings. *See:* Cost analysis *and* Economic analysis
SCBA. *See:* Self-contained breathing apparatus (SCBA)
Schein, E. H., 178
Schermerhorn, J. R., 79, 80, 82–83, 85
SDWA. *See:* Safe Drinking Water Act (SDWA), 1974
Self-contained breathing apparatus (SCBA), 90–91
Sensitivity analysis, 141–142
September 11, 2001, 107
SERC. *See:* State emergency response commission (SERC)
SFPE. *See:* Society of Fire Protection Engineers (SFPE)
SHARP. *See:* Safety and Health Achievement Recognition Program (SHARP)
Shewhart, Walter, 159–160
Shingo, Shigeo, 38
SIC. *See:* Standard Industrial Classification (SIC) Code
Simmons, R., 174
Simple interest, 138, 140
Single payment discount factor, 138
SIPs. *See:* State implementation plans (SIPs)
Situational leadership, 81–82
Social responsibility (SR). *See:* Corporate social responsibility (CSR)
Society of Automotive Engineers (SAE), 181

Society of Fire Protection Engineers (SFPE), 181
Specifications for Quality Programs for the Petroleum, Petrochemical and Natural Gas Industry (API, 2007), 180
Split-half method, 156
SSPP. *See:* System Safety Program Plan (SSPP)
Stakeholders, corporate social responsibility, 110
Standard Industrial Classification (SIC) Code, 40
Star Demonstration Program, 168
Star Program, 168. *See also:* Voluntary Protection Program (VPP)
State emergency response commission (SERC), 10
State implementation plans (SIPs), 8
Statement of work, 69
Statistical process control, 159–165
Stewart, Rodney D., 67
Superfund Amendments and Reauthorization Act (SARA), 1986, 10. *See also:* Comprehensive Environmental Response, Compensation, and Liability Act (CERCLA), 1980
Surveying, assessment, 157
Sustainability *See also:* Continuous improvement process (CIP)
 ISO 26000, 106
 L'Oreal report, 115–116
System, defined, 66
System safety
 defined, 173–174
 framework for, 173–177, 183–184
System Safety Handbook (NASA), 175, 180
System Safety Program Plan (SSPP), 175
Systems Engineering Handbook (NASA), 175

T

TapRooT, 38–39
Task statements, 69–70
Tax benefits, 131–132
Team busters, 83
Teams, building, 82–84
Terrorism, 120–122
Test-retest method, 156
Time study, 95–96
Time value of money, 76–77, 137–139
Total recordable case rate (TRCR), 40–41
Toxic Substances Control Act (TSCA), 1976, 9
Trailing indicators, safety effectiveness, 156
Training
 assessment, 90–92
 education, safety professional, 49–51
 evaluation, 154
 General Duty Clause (OSHA) obligations, 20
 OSHA standards, 36, 153
 safety and health program, 153
 Voluntary Protection Program (OSHA), 22
 workforce, 90–92
Training Requirements in OSHA Standards & Training Guidelines (OSHA), 153

Transparency, business risk, 105–106
Travel, risk and, 120–122
TRCR. *See:* Total recordable case rate (TRCR)
Treatment technique (TT), 8
TSCA. *See:* Toxic Substances Control Act (TSCA)

U

U control charts, 163, 165
Underwriters Laboratory (UL), 4
Union Carbide, 102, 103, 127–128
United Nations Global Compact, 105–106
U.S. Department of Agriculture (USDA), 10
U.S. Department of Defense (DOD), 175–176, 182
U.S. Department of Defense Instruction (DoDI), 180
U.S. Department of Energy (DOE), 3
U.S. Department of Labor (DOL), Bureau of Labor Statistics (BLS), 40, 144
U.S. Department of Transportation (DOT)
 commercial vehicles, 62
 hazardous materials, 3
 history and programs, 3

V

Variable costs, 132
Vincoli, J., 184
Voluntary Protection Program (VPP), 5–7, 22, 167–168

W

WHMIS. *See:* Canadian Workplace Hazardous Materials Information Systems (WHMIS)
Williams-Steiger Act, 5. *See also:* OSHA, OSH Act
Wood, Stuart, 114
Work breakdown structure, 69–70, 147
Work sampling, 92–94
Work-site analysis, 28–30, 152–153
Workers. *See:* Employees
Workforce analysis, 87
Workplace safety. *See: safety entries*

X

X charts, 160–161, 164

Z

Z10 Standard
 risk management, 107
 safety engineering management, 24–26, 34–35, 46–47
 safety management, 17–18, 25–26, 30, 42–43
Z15 Standard, 59–64
Zaleznick, Abraham, 79–80

Printed and bound by PG in the USA